Mathematical Action & Structures of Noticing

Brian Fleming Research & Learning Library
Ministry of Education
Ministry of Training, Colleges & Universities
900 Bay St. 13th Floor, Mowat Block
Toronto, ON M7A 1L2

Mathematical Action & Structures of Noticing

Studies on John Mason's Contribution to Mathematics Education

Stephen Lerman
London South Bank University, UK

Brent Davis
University of British Columbia, Vancouver, Canada

SENSE PUBLISHERS
ROTTERDAM/BOSTON/TAIPEI

A C.I.P. record for this book is available from the Library of Congress.

ISBN 978-94-6091-029-6 (paperback)
ISBN 978-94-6091-030-2 (hardback)
ISBN 978-94-6091-031-9 (e-book)

Published by: Sense Publishers,
P.O. Box 21858, 3001 AW
Rotterdam, The Netherlands
http://www.sensepublishers.com

Printed on acid-free paper

All Rights Reserved © 2009 Sense Publishers

No part of this work may be reproduced, stored in a retrieval system, or transmitted in any form or by any means, electronic, mechanical, photocopying, microfilming, recording or otherwise, without written permission from the publisher, with the exception of any material supplied specifically for the purpose of being entered and executed on a computer system, for exclusive use by the purchaser of the work.

CONTENTS

Preface   vii
   Steve Lerman and Brent Davis

1. Mathematics Education: Theory, Practice & Memories over 50 years   1
   John Mason

Part 1: Thinking Mathematically   15

2. The Development of Mathematical Thinking: Problem-Solving and Proof   19
   David Tall

3. The Excircle Problem: A Case Study in How Mathematics Develops   31
   Derek Holton, Michael Thomas and Anthony Harradine

4. Just Enjoying the Mathematics   49
   Tim Rowland

5. Self-study as a Basis of Prospective Mathematics Teachers' Learning of Problem Solving for Teaching   63
   Olive Chapman

6. Visualisations as Examples of Employing Students' Powers to Generalize   75
   Elke Söbbeke and Heinz Steinbring

Part 2: The Discipline of Noticing: Mathematical Pedagogy and Pedagogic Mathematics   85

7. Towards a Curriculum in Terms of Awareness   89
   Dave Hewitt

8. Abstraction Beyond a Delicate Shift of Attention   101
   Tommy Dreyfus and John Monaghan

9. Gaining Insight into Teaching and Learning Mathematics at University Level through Mason's *Inner Research*   111
   Elena Nardi

10. Breaking the Addition Addiction: Creating the Conditions for Knowing-to Act in Early Algebra   121
   Marta Molina, Rebecca Ambrose, Encarnación Castro and Enrique Castro

11. Towards an Aesthetics of Education   135
   Alf Coles

CONTENTS

12. Spirituality and Student-Generated Examples: Shaping Teaching to Make Space for Learning Mathematics 147
    Laurinda Brown

Part 3: Variation and Mathematical Structure 161

13. Lesson Play: A Vehicle for Multiple Shifts of Attention in Teaching 165
    Rina Zazkis, Nathalie Sinclair and Peter Liljedahl

14. Analyzing and Selecting Tasks for Mathematics Teaching: A Heuristic 179
    Pedro Gómez and María José González

15. Cultures of Generality and Their Associated Pedagogies 189
    David Pimm and Nathalie Sinclair

16. Noticing: Looking Forwards and Backwards 203
    Andy Begg

17. Thinking Mathematically, Disciplined Noticing and Structures of Attention 211
    Anne Watson

Contributors and Acknowledgments 223

Author Index 227

Subject Index 231

STEVE LERMAN AND BRENT DAVIS

PREFACE

This book is not, and at the same time is, a tribute book to the enormous contributions made by our colleague and friend John Mason to the research field of mathematics education, and to mathematical pedagogy, across the world.

It is not a tribute book because every chapter is a report of research and thinking by the authors, not simply a statement of appreciation. Some are accounts of mathematical activity; some are reports of research studies; others are attempts to develop new ideas and perspectives. All engage with how John's ideas have influenced thinking and research and the authors demonstrate how they have taken those ideas forward, or changed them to extend their own research and thinking.

At the same time it is a tribute book because it is a collection of chapters by authors with a specific focus, that of how their work, as presented here, has been inspired by John through his substantial opus and his vibrant presence in a network of mathematics educators.

In the opening chapter, John provides a sense of that body of work, contextualized in a personal account of the evolution of the field of mathematics education over recent decades. Before we go any further, we should mention that this autobiographical chapter on his years of his work in mathematics and mathematics education was actually written by him for 'another purpose'; at least that is what he was told when invited to write that reflective piece. It was always intended to be the opening chapter to this book. Our confidence that no one could craft a better introduction more than outweighed the anxiety associated with the small deception needed to secure the chapter. (That was not our only anxiety as editors. Another was that, to the point of going to press, we seem to have managed to keep this project a secret from John. That was no small feat. As the list of contributing authors reveals, John is not just well known, but well connected.)

John's introductory chapter renders the task of writing a Preface a very simple one, that of explaining how we have structured the book. We chose three themes that have permeated the research that has grown out of his ideas:

Thinking Mathematically

The Discipline of Noticing: Mathematical Pedagogy and Pedagogic Mathematics

Variation and Mathematical Structure.

It's not easy to separate John's work into three, apparently separate ideas, nor has it been easy looking at each of the chapters and allocating them to one of the parts of the book. Like John's own work across these topics, there are overlaps, synergies and resonances across the labels. Nevertheless, we hope we have done justice to the chapters written and to John's work, in our classification.

The book ends with a chapter by Anne Watson. At the beginning we consulted substantially with Anne, but once the outline of the book project had become clear we took over the whole responsibility for it. Anne's chapter looks at her own and John's work as it is currently developing. After all, Anne knows John rather well!

Steve Lerman
London South Bank University (United Kingdom)

Brent Davis
University of British Columbia (Canada)

JOHN MASON

1. MATHEMATICS EDUCATION: THEORY, PRACTICE & MEMORIES OVER 50 YEARS

INTRODUCTION

Having been asked to look back over my life in mathematics education, I take the liberty of recalling some of the most stimulating moments as they come back to me, in an attempt to analyse what mathematics education has been about for me. In particular I want to suggest that while the field has maintained and even widened the gap between theory and practice, it is incumbent upon us to keep in front of us that the purpose of our work is to understand and contribute to student learning of mathematics. One way I have consistently attempted to do this is to try to *preach what I practice* rather than the other way round.

One of my great amusements is that in the UK I am, I believe, seen mainly as a theorist, someone who works with ideas and tries to make them available to others. By contrast, in other countries I believe I am seen as immensely and intensely practical. In my defence, in the UK I can point to numerous publications that have offered teachers practical actions to initiate when working with learners. But as I have often said to staff on arrival at the Open University, you have 12 to 18 months to establish a reputation in the university; after that it is very hard to alter. So too in the academic world more generally. For example, it seems that when people attend one lecture-presentation, they assume that that is all the person can talk about, when actually most of us are willing to engage with a wide range of issues at different phases of mathematics education.

SOME HISTORICAL ACCOUNTS

I say 'memories over 50 years' in the title, because I started tutoring at the age of 15 at the request of my mathematics teacher Geoff Steele. It was only later that I realised just how profoundly he had influenced me through his stimulation and challenge. Many years later still I discovered that he had had no training in teaching nor much in mathematics: he worked to keep just ahead of me, writing out in longhand theorems in projective geometry, leading me through Hall & Knight on continued fractions, and engaging me in Susanne K. Langer's Symbolic Logic. I came to value very highly the contact I had with topics that did not appear in the formal curriculum until late university, and on this I base my recommendations that students be challenged sideways, in breadth and depth, rather than accelerated through the curriculum. It was true that I twice 'skipped' ahead, but this provided space to consolidate and explore broadly, and I am ever so grateful for it.

S. Lerman and B. Davis (eds.), Mathematical Action & Structures of Noticing: Studies on John Mason's Contribution to Mathematics Education, 1–14.
© 2009 Sense Publishers. All rights reserved.

My first attempts at tutoring were of course completely naïve. I explained things when students I was tutoring got stuck. I tutored high school students in my first year at university, and then started supporting students in my college in the years below me. I discovered later that they used to go first to my friend across the hall who knew how to do the problems; then they would come to me and watch me struggle, asking them questions about what theorems they knew and so on! In my third and fourth years I tutored for the university. It was here that I discovered the effectiveness of *being mathematical with and in front of learners*, although I didn't formulate this slogan until much later. I would face a class of students stuck on a problem about which I knew almost nothing. So I would ask them to read the problem out loud, then to tell me from their notes what the technical terms meant, and what theorems they had in their notes concerning them. In every case I eventually 'saw' how to resolve the problem (and to my retrospective regret) I would then show them how to do the problem. At least they saw a slightly more experienced learner struggling publicly, so they could pick up some practices for themselves.

In graduate school I was shown George Pólya's (1965) film *Let Us Teach Guessing* on the Friday afternoon before teaching a semester class that began at 7:45 each morning, starting on the Monday. As I later realised, the film released in me many of the practices used with me by Geoff Steele, and resonated vibrantly with practices I had developed spontaneously so as not to look a complete fool in tutorials. I got agreement with the class that they would work my way on Mondays; on Tuesdays to Thursdays we would then 'finish the chapter of the week', and on Fridays I would do revision problems with and for them. Within a week or two we were working 'my way' until at least part way through Wednesdays, finishing the chapter Thursdays, and revising on Fridays.

So what did it mean to 'work my way'? My recollection is that I would ask them questions to get them thinking. I would construct examples and generally cajole them. I would summarise in technical language what I thought they had begun to 'see' or appreciate. I felt that I was engaging them in the thinking. I am sure that an observer would have seen me getting them to 'guess what was in my mind', but the class was interactive and I hope challenging.

On my arrival at the Open University I realised that distance education was the antithesis of what I thought learning was about, but I settled down to write material. It was then that I realised how problematic teaching mathematics is, and that I had been enculturated as a structuralist through the influence of Bourbaki on many of my lecturers. One advantage was that, unlike the practices in a face-to-face institution where each lecture is an event to be survived, I frequently found myself at my desk wondering what example to use, what definition to use, what theorems to put in what order, and how best to interlace examples with abstractions and generalities.

As I was one of the few people in the faculty who had worked with Pólya's ideas, I was asked to organise the summer schools. I instituted sessions such as *investigations* (inspired by what I discovered going on in primary schools in the UK and based on Pólya's film), *mental callisthenics* (reproducing sessions I had experienced myself in primary school), *surgeries* (where students could come and ask for help on any aspect of the course), *tutor revelations* (in which a tutor would work through some typical questions while exposing the inner thoughts, choices and incantations accompanying the solution) as well as lectures and tutorials. I even tried sessions called *tutor bashing* in which a tutor poses a question to another tutor who then works at it from cold in front of the students before posing a question to the next tutor in sequence. This was an attempt to show students that tutors were mortal and fallible and that mathematics does not flow perfectly out of the pen in its completed form.

It took me several years to realise that what was obvious to me about how the mathematical practices described by Pólya, such as specialising & generalising, imagining & expressing, conjecturing & convincing, were not obvious to many of our tutors. Thus some tutors adopted a "don't let them leave for coffee break without giving them the formula, because they'll just get it from others anyway" approach, and recommended this to tutors new to the summer schools. Furthermore on investigation it transpired that students in tutorials where the tutor did not like investigations usually came out not liking them, and students with tutors who did like them usually came out enjoying them. The stance, beliefs and attitudes of the tutor could be highly influential. This later resonated with the pioneering work of Alba Thompson (1984, 1992) working with teachers.

In 1973–74 I spent nearly a year with some 125 others in a house in Gloucestershire under the direction of J.G. Bennett, mathematician, scientist, linguist and seeker. It was a pivotal year for me, crystallizing many awarenesses and awakening me to many others. Ever since then I have been reconstructing the ideas and practices I encountered. In particular, I experienced deeply an experiential approach to enquiry, which for me extended through mathematics to every aspect of life and thought. Many years later I decided to devote time to reconstructing those practices expressed in the domain of mathematics education in particular, but applicable in any caring profession. I called it the Discipline of Noticing (Bennett, 1976; Mason, 1996, 2002). My idea was to provide teachers and other carers with a philosophically well-founded method and theoretical framework for researching their own practice, that is, for working on themselves.

I began my research life in combinatorial geometry, and achieved a certain notoriety within the rather small community of like-minded scholars as someone with a practical, example-rich approach to tackling really difficult mathematical problems from a structural perspective. It was after an event run by a colleague, Johnny Baker, in which teachers from secondary mathematics and science departments reported on their experience of teaching and using mathematics in schools, that I realised that although the mathematical problems I worked on were difficult, very few people cared, whereas the problems in mathematics education are essentially unsolvable, but a very large number of people care. So I turned my attention explicitly to mathematics education, and set myself three years to establish a reputation in the field. However, I have never lost my interest in, no, my addiction to, working on mathematics. I always have some mathematical problem in the back of my mind that I work on in otherwise idle moments. This serves to keep me in touch with my own experience and sensitises me to struggles that others may have with different topics, concepts or problems.

It was while designing a course for mathematics teachers that I persuaded my colleagues that it would be a good idea for teachers to engage in mathematical thinking for themselves at their own level each week. But then a decision had to be made as to how to choose what problems to offer them, and how to structure that experience. We found that we had plenty of advice to offer, in my case gleaned from Pólya, Bennett and awareness of my own experience. Together Leone Burton and I decided to write a book about problem solving, in order to assist us in developing a structural framework for making suitable selections for the course. Leone introduced me to Kaye Stacey, and *Thinking Mathematically* (Mason et al., 1982) was born. Unsurprisingly, perhaps, I never had any opportunity to teach a course from it, though others have and continue to do so more than 25 years on.

My non-academic activities in the 70s brought me into contact with a wide range of group activities exploring sensitivity training and body-mind connections. I encountered a wide range of practices and authors such as Abraham Maslow (1971). I immediately recognised from him that I was really interested in what is possible, what *could* be, rather than *what is the case currently*. Since at the Open University there were no students on campus to use as subjects, nor easy contact with schools, it matched my situation to be more interested in what is possible through examining my own experience while also working with teachers in in-service professional development sessions. One effect was that, with only a few exceptions, I only ever worked with teachers and students on a one-off basis rather than over an extended period of time.

Soon after I arrived at the Open University I discovered the Association of Teachers of Mathematics, and began to go to their meetings. I encountered there a group working on mental imagery and this fired my imagination, literally. I adopted and adapted practices that Dick Tahta and others used with posters and animations, to use with videotape of classroom interactions, and this was the basis for what our group constructed as an approach to using video. At ICME 5 in Adelaide I discovered this was called or at least akin to constructivism. Combined with my experiential orientation, this set me up to resonate strongly with Ernst von Glasersfeld when I met him in Montréal in 1984.

In Adelaide I also met Guy Brousseau and extended contact with Nichol Balacheff. Impressed by several of the constructs they used, in particular *transposition didactique* (Chevallard, 1985), *situation didactique* (Brousseau, 1984, 1997) and the *didactic tension*, and *epistemological obstacles* (Bachelard, 1938), it took me a long time to appreciate even the most surface features of the deeply analytic frame that informed their impressive research. However, my own experience in working with groups of teachers experientially had shown me that offering results of research enquiries to others is in itself highly problematic.

When I was a graduate student there were frequent calls for research on the most effective way to teach mathematics to undergraduates: 3 lectures + 2 tutorials per week, or 5 lectures per week, or 3 tutorials and 2 lectures or what? It was evident to me that the issue depended on too many factors connected with the setting, the individuals, the expectations, and the practices within lecturing and tutorials to be able to declare one better than another universally. Whereas most mathematicians that I knew were seeking a mathematical-type of theorem with definitive conclusions, I was convinced that any value system would be situation dependent. I found statistical findings deeply unsatisfying, because either they would agree with my own prejudices, in which case they told me nothing, or they would contradict those biases, in which case I would reject them as being unsuitable or irrelevant to my situation. I felt perfectly at home with the impossibility of mathematical-like theorems in mathematics education, because of the presence of human will, intention and ideals.

When I started working with teachers in the UK I soon realised that making use of 'findings' was problematic for others as well. Proposing a research finding (qualitative or quantitative) that is close to current practice is likely to be responded to by assimilation without noticing any subtle differences. Proposing a research finding that challenges thinking or which is not immediately compatible with practices, is unlikely to stimulate people to try the idea much less adopt it simply because it is a research finding. There has to be something that catches attention, either because it seems implausible and so motivates checking, or because it appears to match perceived current needs.

In my own case, even if I do try something, I am most likely to modify it to fit with my perspective and approach. Indeed, there is no 'it' as such, only my re-construction. This corresponds with my view of classroom incidents, indeed incidents and events generally: there is no 'event' as such, merely the stories told about it, whether at first, second, or later hand. On the other hand, if the finding fits with my experience, it is likely to seem 'obvious', so I am likely to pay no attention to any slight differences that might in fact be significant. Instead, I feel reassured and carry on. Thus for me a successful professional development session is one in which participants can actually imagine themselves acting differently in some situation in the future which they recognise. This statement is much more significant than it may appear at first sight. I emphasise *imagine themselves*, for what I learned from Bennett was the immense power of mental imagery for preparing actions to take place in the future.

What does seem to be helpful is prompting people to experience something which sheds light on their past experience and offers to inform their future choices. I see professional development as personal enquiry, stimulated and supported by work with colleagues, but essentially a psychological issue with a socio-cultural ecology. I have however always resisted pushing this as far, for example, as my one-time colleague Barbara Jaworski (2003, 2006, 2007) has done. I am content to indicate possibilities to others rather than trying to maximise efficiency and efficacy. For me, change is such a delicate matter that it must be left to individuals within their various communities, to the extent that "I cannot change others; I can work on changing myself", and even that is far from easy!

My interests have always been in supporting others in fostering and sustaining mathematical thinking in their students. I have at various times concentrated on mental imagery, modelling, problem solving, and language, but these have been but byways in getting to grips with the nature and role of attention. I have, for example, found it convenient to shift my discourse from *processes of problem solving* to *exploiting natural powers*, finding that the same ideas (imagining & expressing, specialising & generalising, conjecturing & convincing, among others) continue to be potent as long as they are expressed in a context which resonates with people's experience and a discourse that they recognise. Early in my reading of mathematics education books and papers I recognised that each generation has to re-express insights in their own vernacular, even though these insights have been expressed before. Indeed, each person has to re-experience and re-construct for her- or himself. This contrasts with mathematics in which it is possible to be directed along a 'highway' towards problems at the boundary without traversing all of the country in between. Mathematics education is not like that, and perhaps never will be, at least until we establish common ways of working. I take up this theme in the next section.

In the early 1980s I had the chance to attend a number of seminars led by Caleb Gattegno when he tried to re-vivify his *science of education* (Gattegno, 1987, 1990) in the mathematics education community in England. I found his approach attractive, with a good deal in common with what I had learned from Bennett, but leading to a rather different cosmology. I began to get a taste of what it is like when an experienced 'grey-beard' assembles their to-them-coherent-and-comprehensive framework or theory. Whereas when the fragments were being worked on and described there is often considerable interest amongst colleagues; once the whole is assembled, people don't really want to know. I ran into this phenomenon again when reading Richard Skemp's later book (1979), where again my experience was one of interest in some of his distinctions, without appreciating all of them, or the way they all fit together. Reading Jean Piaget, Zoltan Dienes, Hans Freudenthal, David Ausubel, Frédérique Papy and Humberto Maturana all had similar effects on me, partly perhaps, because I came across their work after or near the end of their careers. I found many specific distinctions of great value, but resisted taking on board their over-arching theories.

I assume that the issue is one of subordination. Philosophers are trained to suppress their own thinking in order to 'think like' the philosophers they are studying. In mathematics education, the intention is to improve the experience of learners, and this pragmatic dimension may contribute to a reluctance to let go of one's own stance in order to enter, absorb and fully appreciate the stance of someone else. Several French colleagues have given me the impression that in France they are more used to subordinating to an established theory, whereas Northern European, anglo-saxon cultures appear to be more pragmatic and less theory-oriented in assembling their own personal framework or theory that 'works for them'.

THE RISE OF MATHEMATICS EDUCATION

Others are more scholarly at researching the ebb and flow, the waxing and waning of salient constructs in mathematics education. My memory is that in the 1970s and early 1980s research interest focused on students. I was, naturally, caught up in the Pólya-inspired *problem solving* discourse of *processes*, as manifested in *Thinking*

Mathematically. As data accumulated, attention turned to student errors and misconceptions. I recall the breath of fresh air when Douglas McLeod and Verna Adams (1989) edited a book on affect and problem solving, and Alba Thompson championed devoting attention to teachers' beliefs as influencing both how people teach, and what students learn. As I look back now, it seems to me that one of the reasons for each generation revisiting and re-constructing classic insights and awareness is that, as well as participating in a process of personal re-construction, each generation finds itself dissatisfied with the explanatory and-or remedy power of the current discourse and foci. The discourse is drained of its power to inform choices? Each generation seeks fresh fields for explanation as to why students, on the whole, do not learn mathematics effectively or efficiently, and latterly, why teachers do not teach what they know and why they know so little of what it is necessary to know in order to teach effectively.

On the one hand we have Henri Poincaré's position of being mystified as to why perfectly rational people can fail to succeed at the perfectly rational discipline of mathematics, and on the other hand we have generations of students convinced that either fractions or algebra was a watershed of their involvement in mathematics. Clearly rationality is not the central feature of most people's psyche. One of the many things that has impressed me about Open University students over the years is that when I used to ask students on our mathematics courses why they were going to all that effort, I almost always got the reply "always liked mathematics at school; never could do it, mind you, but always wanted to know more". Something touches people, even if it remains dormant.

Attention in mathematics education research has shifted variously between the structure of and inherent obstacles in specific topics, psychological aspects of learning mathematics, psychological aspects of teaching mathematics, sociological aspects of teaching and learning mathematics, acts of teaching, teachers' beliefs and how they influence learners, the historical-socio-cultural forces at work in and through institutions, and the content and format of teacher education courses, not to say the obstacles encountered by novice teachers due to weak mathematical background, and the destructive forces of school practices and government policies on the ideals and aspirations of novice teachers emerging from teacher education courses, to name but a few. Most of what is accepted as *research* involves making observations of others (what I call *extra-spective*) whether associated with deliberate interventions or not. Observations and transcripts are turned into data by being selected for analysis. Analysis then applies a framework for making distinctions, or generates or modifies such a framework.

It is tempting to say that we (the community of mathematics educators, scholars and researchers) have accumulated a great deal of data. We have individually, though perhaps not collectively, drawn a multitude of distinctions and formulated a plethora of constructs to analyse and account for what has been observed. But what have we really got to show for all this effort? Publications proliferate faster than I for one can read them, much less take them in and integrate them. Rarely do we get evidence that the framework has enabled teachers to modify their practice and so influence student learning. So what is the mathematics education enterprise?

THE ENTERPRISE OF MATHEMATICS EDUCATION

On the surface, it is reasonable to expect that those engaged in mathematics education research and scholarship have as their aim the improvement of conditions for learning (and hence for teaching) mathematics more effectively, at every age and stage. Consequently evidence of effectiveness must lie ultimately in improvement in learner experience and performance, in both the short term and long term.

Of course there is an immediate obstacle, for there is little or no agreement as to what constitutes evidence of learner experience, much less evidence of improvement. Since it is easiest to gauge by scores on tests, national and international studies administer tests and pronounce on the results. Questionnaires and even interviews with selected subjects can be carried out. But there is a fundamental difficulty. Test results only indicate what subjects did on one occasion under one set of circumstances with or without specific training in preparation. Interviews at best reveal only what the interviewer probes and selects due to their sensitivities, and questionnaire responses are highly dubious indicators of what lies beneath surface reactions to specific questions. As soon as you identify an indicator of mathematical thinking or other mathematical competence or success, it should take a competent teacher at most two years to work out how to train students to answer those 'types' of questions. It all comes back to Guy Brousseau's notion of the *didactic contract* as manifested by the *didactic tension* and its parallel, the *assessment tension*:

> The more clearly and specifically the teacher (assessor) indicates the behaviour sought, the easier it is for the learner to display that behaviour without generating it from themselves (understanding).

Put another way in a discourse derived from Caleb Gattegno, training behaviour is important and useful, but it is inflexible and even dangerous if it is not paralleled with educating awareness. It is awareness (what enables you to act, what you find 'comes to mind' in the way of actions) that guides and directs (en)action, using the energy arising from affect (by harnessing emotions). It is ever so tempting to train someone's behaviour by giving them rules and mnemonics to memorise, and quantities of exercises on which to rehearse. But real learning only occurs when these form the basis for reflection and integration so that awareness is educated (as in the Confucian culture approach to teaching and learning). Alternatively, one can work on educating awareness and training behaviour together, through harnessing emotion, and this is the approach that I have endeavoured to practice, and through practicing, to articulate for myself and others so as to make the process more efficient over time. Because I am interested in what is possible, and because the only way of directing other people's attention is through being aware of the focus of my own attention, I use *intra-spection* (between selves or between people) as distinct from *intro-spection* which got a negative connotation through the indulgences of people trying to develop phenomenological research methods in the early part of the 20th century.

One of the underlying tensions in mathematics education that I am aware of is that between a 'scientific stance' and a 'phenomenological stance'. Editors want their journals to contribute to the scientific development of knowledge. Journals have recently become so obsessed with theoretical frameworks that papers get longer and longer, without any growth in substance. I suspect that colleagues, especially editors, want to see mathematics education build a coherent and well-founded structure of knowledge. They like to see people building on each others' work, adding to and refining rather than starting afresh. I wonder however whether this is even possible, much less desirable, given the nature of focus of mathematics education, working as it does with human beings placed in institutional settings of various sorts, and exercising their wills and intentions through their dominant dispositions. I want to put a different case, the case of working with lived experience.

STRUCTURED AWARENESS

I have often thought and sometimes said, that when I am engaged in my enquiries, I enjoy it most when I am at the overlap between mathematics, psychology and sociology, philosophy and religion. There is something about working on a mathematical problem which is for me profoundly spiritual; something about working on teaching and learning that integrates all three traditional aspects of my psyche (awareness, behaviour and emotion, or more formally, cognition, enaction and affect) as well as will and intention, which themselves derive from ancient psycho-religious philosophies such as expressed in the Upanishads (Radhakrishnan, 1953) and the Bhagavad Gita (Mascaró, 1962; see also Raymond, 1972). I associate this sense of integration with an enhanced awareness, a sense of harmony and unity, a taste of freedom, which is in stark contrast to the habit and mechanicality of much of my existence. Even a little taste of freedom arising in a moment of participating in a choice, of responding freshly rather than reacting habitually is worth striving for.

One way to summarise such experiences is that, in the end, what I learn most about, is myself. This observation is not as solipsistic, isolating and idiosyncratic as it might seem, for in order to learn about myself I need to engage with others (who may, as is the case for hermits, be virtual), and I need to be supported and sustained in those enquiries. A suitable community can be invaluable, though an unsuitable community can be a millstone! I reached this conclusion through realising that when a researcher is reporting their data, and then analysing it, the distinctions they make, the relationships they notice, the properties they abstract all tell me as much about their own sensitivities to notice and dispositions to act as they do about the situation-data being analysed. Indeed I proposed an analogy to the Heisenberg principle in physics: the ratio of the precision of detail of analysis to the precision of detail about the researcher is roughly constant (Mason, 2002, p. 181).

The seeds of this observation were in working with teachers on classroom video, informed by techniques for working on mathematical animations (e.g. those of Jean Nicolet and Gattegno's reworking of them). The technique is to get participants to reconstruct as much of the film as they can after seeing it just once. When they have made a good attempt, they have specific questions about portions they only partly recall, or where different people have different stories. So a second viewing makes sense, but only because there are specific questions. Applied to classroom video, we adopted a similar stance in order to counteract the common reaction of "I wouldn't let that teacher in my classroom", or "my low attainers are lower attaining than those low attainers" (Jaworski, 1989; Pimm, 1993). It seemed that teachers saw classroom video as a challenge to their identity and practices. By getting them to recount specific incidents briefly but vividly from the video with a minimum of judgement, evaluation, explaining, we found that they soon recognised incidents as being similar to incidents they had met in their own experience. So the videos became an entry into participants' own past experience, and hence gave access to their lived experience. This makes so much more sense than critiquing the behaviour of some unknown teacher whose class may already have left school, so there is no way that their behaviour could be altered!

Incidents that strike a viewer usually resonate or trigger associations with incidents recalled from the past. Describing these to others briefly-but-vividly so as to resonate or trigger their own recollections provides a database of rich experiences which can be accessed through the use of pertinent labels. Often sameness and difference between re-constructed incidents has to be negotiated amongst colleagues, and this is what prompts probing beneath the surface. As Italo Calvino (1983, p. 55) said, "It is only after you come to know the surface of things that you venture to see what is underneath; but the surface is inexhaustible".

I have come to recognise that Bennett (among many others over the centuries) was right when he highlighted the fundamental act of making distinctions. It is, after all, how organisms at all levels of complexity operate. Change is the experience of making distinctions over time; difference is the experience of distinction making in time. Evaluation is the experience of distinguishing relative intensities (as a ratio or scaling). Bennett went much further, amplifying Gurdjieff's observation that 'man is third force blind' (see Orage, 1930). In other words, distinguishing things, this from that, is important, but locks you into tension or evaluation, and is just the beginning of what is possible. In order to appreciate how the world works (whether material, mental, symbolic or spiritual) it is necessary to become aware that actions require three impulses, something to initiate, something to respond, and something to mediate between these, to bring them into or hold them in relationship. The product of actions can then go on to serve to initiate, respond to, or mediate a further action. Bennett continued this neo-pythagorean analysis into the quality of numbers from 1 to 12 in his monumental four-volume work *The Dramatic Universe*, which he called *systematics*, long before 'systems theory' became a slogan. Perhaps because of my structural upbringing, I found myself resonating with his approach, to the extent that I could sometimes hear in the structure of his talks the ways in which he was systematically employing systemic qualities of a particular number.

PRECISION & REPLICATION

There is another issue concerning transforming observations into data and the degree of precision presented in research reports. Over the years there has been an evident growth in the length and complexity of papers in mathematics education. It used to be that some detailed transcripts along with some analysis stimulated colleagues to investigate the phenomenon in their own setting. A classic example would be the paper by Stanley Erlwanger (1973, reprinted in Cooney *et al.*, 2004) about Benny's encounter with fractions. Nowadays this paper would probably be rejected by journals as failing to present an adequate theoretical framework and discussion of method and ethics. Many papers are so heavily theory-laden in the opening sections that by the time I get to the substance I have forgotten exactly which parts of which theory are actually being employed, and indeed sometimes it is not even very easy to detect this. It seems to me that often only tiny fragments of theoretical frameworks are called upon. Indeed, I have no problem with this at all, because of my eclectically cherry-picking approach to understanding and practice: all I can ever do is be stimulated or sensitised to notice, that is to discern details not previously attended to, and through that discernment, raise questions deserving of enquiry. But if authors are selecting fragments, why not be straightforward about it? I go so far as to suggest that experience itself is fragmentary, despite consciousness and the collection of selves that make up consciousness and personality trying to develop stories to make it look continuous and coherent (Mason, 1986, 1988). This is the one detail on which I disagree with William James' notion of a 'stream of consciousness'. My own observations agree with Tor Nørretranders (1998) that these stories are a fabricated illusion.

I realise that editors have a commitment to building the scientific foundations of mathematics education, but I don't see that present practices are actually furthering the field, in the main. What we do have is a plethora of distinctions, sometimes several labels for at best subtly distinct distinctions, and sometimes the same label is used for different distinctions. What the field really needs is some agreement on ways of working, rather than on theoretical frames and stances. We need to build up a vocabulary for how we compare observations, turn them into data, and negotiate meaning amongst ourselves. This would then make it easier to offer similar distinctions to others including teachers, teacher educators and policy makers, and to negotiate similarities, differences and intensities. Caleb Gattegno offered his *science of education* but this is too radical for most to agree to; in the Discipline of Noticing I tried to offer a less radical and more practical foundation for ways of working; I am sure that others feel they have done the same. The problem in my view lies not in the fact that everyone discerns slightly differently, but that we don't have established ways of negotiating similarities and differences in what is noticed and in what triggers that noticing, and in what actions might then be called into play.

Despite the developments in style (I hesitate to use the word *improvement*) it is still rarely if ever possible to imagine much less actually carry out a replication of a study reported on in a mathematics education research paper in a journal. There is simply never enough detail. I happen to suspect that it would never be possible to replicate a study exactly, precisely because of the complex range of factors comprising the traditional triad of student, teacher and mathematics all embedded in an institutional environment.

If it is either impossible or not necessary to be able to replicate the conditions of a study, what is it that we are gaining by reporting on our studies? My radical response to such a question is that what matters most is *educating awareness* by alerting me to something worth noticing because it then opens the way to choosing to respond rather than react with a more creative action than would otherwise be the case. I don't need all sorts of detailed data, because the more precise and fine-grained the detail, the less likely I am to pay attention to the over all phenomenon being instantiated, and so the less likely I am to recognise it again in the future and so choose to act differently.

REPRISE

I am genuinely perplexed about the role and nature of structure in a domain such as mathematics education. On the one hand, with my structural background, I find it really helpful to be able occasionally to invoke one or other structure in order to inform my thinking. But I have colleagues who resist such an approach, just as I have resisted accommodating the whole of other people's structured frameworks. It is too simplistic to say that each could be expounded and then tested by experiment to see which is 'best'. I am reminded of a sequence of lectures I was required to attend in my first year at university on *the leap of faith*. My recollection is that they were about the philosophical conundrum of how you cannot investigate or enquire into what it is like to believe something without actually believing it. Put in an overly extreme form perhaps, 'if you can critique it, you haven't experienced it fully'. Of course this is anathema to many in Western society, but increasingly popular to fundamentalists the world over.

My own experience is that I do not usually use my own frameworks systematically or mechanically, because they have been integrated into how I perceive the world and how my thinking progresses. Every so often it is useful to ask myself if I have taken all aspects of an action, an activity, a potentiality, a moment, a transformation into account, and this is when it can be fruitful to remind myself of the pertinent number and its structural qualities. More specifically, the Structure of a Topic framework (Griffin & Gates, 1989; Mason & Johnston-Wilder, 2004a, 2004b), based on Bennett's 'present moment' system associated with qualities of six is particularly useful when preparing to teach a topic. The six modes of interaction (expounding, explaining, exploring, examining, exercising and expressing) arising from the qualities of three alert me to possible forms of interaction and bring different interventions to mind (Mason, 1979).

I suspect that each of us does something similar. We act in the world; when some tension or disturbance arises, we resort to accustomed modes of thinking using whatever frameworks come to mind, and then carry on. We have habitual but slowly evolving forms of activity with which we feel comfortable; we are sensitised to certain aspects of potentiality in a situation; we stress certain aspects of the present moment; and so on. Calvino (1983, p. 107) said something similar: "The universe is a mirror in which we can contemplate only what we have learned to know in ourselves", which in turn resonates with a North American shaman Hyemeyohsts Storm (1985) who phrased it as "the Universe is the Mirror of the People". The universe, whether material, imagined or symbolic, provides a mirror for seeing ourselves and so bringing possible actions to mind. Indeed, all we can see is in fact ourselves, in the sense that what we discern and relate is a reflection of ourselves. What professional development means is ongoing work to extend sensitivities, striving for a greater balance in the interplay of component features, so as to participate more fully in the evolution of awareness.

REFERENCES

Bachelard, G. (1938/1980). *La formation de l'esprit scientifique*. Paris: J. Vrin.
Bennett, J. (1956–1966). *The dramatic universe* (Vol. 1–4). London: Hodder & Stoughton.
Bennett, J. (1976). *Noticing. Volume 2 of the sherborne theme talks series*. Sherborne, UK: Coombe Springs Press.
Brousseau, G. (1984). The crucial role of the didactical contract in the analysis and construction of situations in teaching and learning mathematics. In H. Steiner (Ed.), *Theory of mathematics education* (pp. 110–119). Bielefeld, DE: Institut fur Didaktik der Mathematik der Universitat Bielefeld.
Brousseau, G. (1997). *Theory of didactical situations in mathematics: Didactiques des mathématiques, 1970–1990* (N. Balacheff, M. Cooper, R. Sutherland, & V. Warfield, Trans.). Dordrecht, NL: Kluwer.
Calvino, I. (1983). *Mr Palomar*. London: Harcourt, Brace & Jovanovich.
Carpenter, T., Dossey, J., & Koehler, J. (Eds.). (2004). *Classics in mathematics education research*. Reston, VA: NCTM.
Chevallard, Y. (1985). *La transposition didactique*. Grenoble, FR: La Pensée Sauvage.
Erlwanger, S. (1973). Benny's conception of rules and answers in IPI Mathematics. *Journal of Mathematical Behaviour*, 1(2), 7–26.
Gattegno, C. (1987). *The science of education, part 1: Theoretical considerations*. New York: Educational Solutions.
Gattegno, C. (1990). *The science of education*. New York: Educational Solutions.
Griffin, P., & Gates, P. (1989). *Project Mathematics UPDATE: PM753A,B,C,D, Preparing to teach angle, equations, ratio and probability*. Milton Keynes, UK: Open University.
Jaworski, B. (2006). Theory and practice in mathematics teaching development: Critical inquiry as a mode of learning in teaching. *Journal of Mathematics Teacher Education*, 9(2), 187–211.
Jaworski, B. (1989). *Using classroom videotape to develop your reaching*. Milton Keynes, UK: Centre for Mathematics Education, Open University.
Jaworski, B. (2003). Research practice into/influencing mathematics teaching and learning development: Towards a theoretical framework based on co-learning partnerships. *Educational Studies in Mathematics*, 54(2–3), 249–282.
Jaworski, B. (2007). Developmental research in mathematics teaching and learning. Developing learning communities based on inquiry and design. In P. Liljedahl (Ed.), *Canadian mathematics education study group proceedings, 2006 annual meeting* (pp. 3–16). Burnaby, BC: CMESG.
Mascaró, J. (Trans.). (1962). *The bhagavad gita*. Harmondsworth, UK: Penguin.
Maslow, A. (1971). *The farther reaches of human nature*. New York: Viking Press.
Mason, J. (1979, Feb). Which medium, which message. *Visual Education*, 29–33.
Mason, J. (1986). Probes and fragments. *For the Learning of Mathematics*, 6(2), 42–46.

Mason, J. (1988). Fragments: The implications for teachers, learners and media users/researchers of personal construal and fragmentary recollection of aural and visual messages. *Instructional Science, 17*, 195–218.

Mason, J. (1996). *Personal enquiry: Moving from concern towards research*. Milton Keynes, UK: Open University.

Mason, J. (2002). *Researching your own practice: The discipline of noticing*. London: RoutledgeFalmer.

Mason, J., Burton, L., & Stacey, K. (1982). *Thinking mathematically*. London: Addison Wesley.

Mason, J., & Johnston-Wilder, S. (2004a). *Designing and using mathematical tasks*. Milton Keynes, UK: Open University. (2006 reprint: St. Albans, UK: QED).

Mason, J., & Johnston-Wilder, S. (2004b). *Fundamental constructs in mathematics education*. London: RoutledgeFalmer.

McLeod, D., & Adams, V. (Eds.). (1989). *Affect and mathematical problem solving: A new perspective*. New York: Springer-Verlag.

Nørretranders, T. (1998). *The user illusion: Cutting consciousness down to size* (J. Sydenham, Trans.). London: Allen Lane.

Orage, A. (1954). *Essays & aphorisms*. New York: Janus.

Pimm, D. (1993). From should to could: Reflections on possibilities of mathematics teacher education. *For the Learning of Mathematics, 13*(2), 27–32.

Pólya, G. (1962). *Mathematical discovery: On understanding, learning, and teaching problem solving* (combined ed.). New York: Wiley.

Pólya, G. (1965). *Let us teach guessing* [film]. Washington, DC: Mathematical Association of America.

Raymond, L. (1972). *To live within*. London: George Allen & Unwin.

Radhakrishnan, S. (1953). *The principal upanishads*. London: George Allen & Unwin.

Skemp, R. (1979). *Intelligence, learning and action*. Chichester, UK: Wiley.

Storm, H. (1985). *Seven arrows*. New York: Ballantine.

Thompson, A. (1984). The relationship of teachers' conceptions of mathematics and mathematics teaching to instructional practice. *Educational Studies in Mathematics, 15*, 105–127.

Thompson, A. (1992). Teachers' beliefs and conceptions: A synthesis of the research. In D. Grouws (Ed.), *Handbook of research in mathematics teaching and learning* (pp. 127–146). New York: MacMillan.

John Mason
Open University & University of Oxford (United Kingdom)

Mathematical
Action &
Structures
Of
Noticing

Part 1:

THINKING MATHEMATICALLY

Part 1:

THINKING MATHEMATICALLY

Easily the best known and most widely distributed of John's publications is his 1982 *Thinking Mathematically* (authored with Leone Burton and Kaye Stacey). We borrow not just the title of that book for this section, but its core theme.

Central to John's approach is that teaching and learning is about the development of mathematical thinking, not on acquiring mathematical knowledge *per se*. Indeed such knowledge is inert without it having been acquired through the challenge to think mathematically. The teacher's role is to support pupils to attend to the powers with which they are born, powers to discriminate between and to see similarity across objects, to conjecture, to inquire, and so on, and to develop those powers in the appropriate directions. For teachers to know what is meant by 'appropriate directions' requires them, indeed all of us involved in the field of mathematics education, to continue to experience those challenges to think mathematically and to sharpen our awareness of the range of features of mathematical thinking that John and his many colleagues have worked on over the years. John will often begin a talk with asking the audience to engage in a mathematical activity first, so that the points he is going to make in his talk are immediate for listeners.

This section begins with a chapter by David Tall in which he incorporates the theory of mathematical thinking into his own theory of the long-term development of the individual that includes the notion of proof. He relates this to Richard Skemp's work on mathematical knowledge and emotions.

The following two chapters, one by Derek Holton, Michael Thomas and Anthony Harradine, and the other by Tim Rowland, give us rich examples of mathematical activity through examples, emphasising and illustrating those features of mathematical thinking that have been described in John's writing.

The fourth chapter in this section, by Olive Chapman, describes a particular research study, drawn from a programme of research over many years, that of pre-service mathematics teachers learning about productive problem solving by their students through their own developing understanding of problems, the problem-solving process, problem-solving pedagogy and problem solving as inquiry-based teaching.

The final chapter in this part is by Elke Söbbeke and Heinz Steinbring, in which they argue that the main goal of mathematics education is to build an understanding of mathematical relations and structures and to practice mathematics on an elementary level, and to achieve this calls for a focus on the abstract and the general right from the start of children's mathematics education. They exemplify their argument through rich examples of young children's mathematical activity.

DAVID TALL

2. THE DEVELOPMENT OF MATHEMATICAL THINKING

Problem-Solving and Proof

INTRODUCTION

It was my privilege to use the book *Thinking Mathematically* (Mason, Burton & Stacey, 1982) for over a quarter of a century from its first publication in 1982 to my retirement in 2007. This was a life-changing experience. Before my encounter with this remarkable text I saw my objective as a mathematics educator to reflect on mathematical knowledge and present it to students in ways that would enable them to make sense of it. In my early career, I wrote books and course notes with this purpose in mind. On the publication of *Thinking Mathematically*, I chose to use the text as a course book for a course that I termed 'Problem Solving' for second and third year undergraduate mathematicians with a liberal sprinkling of computer scientists, mathematical physicists and others.

I remember my abject fear when I first met with these students. I was going to start with the first problem in the book, inviting the students to work out whether it was better to calculate a percentage discount before or after adding a percentage tax. My panic was noted by my secretary in those early days as I walked by her door looking nervous and she said, 'You're doing that problem-solving again, aren't you?'

My fear arose because these were very able mathematics students and it was quite likely that they would say, 'but you just multiply the two factors and multiplication is commutative.' But none of them did.

Place someone in an unusual context and present him or her with a problem and it is likely that they will initially lose all sense of direction and need to build up their confidence. This happened to me and it happened to my students. Over time we developed confidence and an ability to anticipate what would happen. It turned the routine learning (or mis-learning) of mathematics into a dynamic act of self-construction and gave most of us concerned a deep sense of pleasure.

Each week we had a two-hour problem solving session with a class of forty to eighty students where I began by setting the scene with the objective of the class, using successive sections of the book then leaving the students to solve a particular problem illustrating the objective of the day. I also announced a 'problem of the week' for students who finished the problem of the day to keep them occupied. Initially some competitive students would move on to the problem of the week fairly quickly, but often they hadn't solved the problem at all.

S. Lerman and B. Davis (eds.), *Mathematical Action & Structures of Noticing: Studies on John Mason's Contribution to Mathematics Education,* 19–29.
© 2009 Sense Publishers. All rights reserved.

The book suggested three levels of explanation:

> convince yourself,
> convince a friend, and
> convince an enemy.

Often the students had a story that clearly convinced themselves and even convinced their friends in the group, but by acting as an enemy I was able to begin to help them be more reflective about what they claimed, so that, over time, they began to question their ideas as a matter of course.

It was my belief that I should not try to solve the problems in advance. It was a distinct advantage to be caring but non-directive in my relationships with the students. *Not* knowing the 'answer' meant that I could change my approach from someone who shows how to do things and gives hints into someone who encourages the students to think for themselves. 'Are you sure?' 'What does this tell you?' 'Is there another way of looking at it?'

At the same time I introduced the students to Richard Skemp's theories of modes of building and testing and, more importantly, to his ideas of goals and anti-goals, to help the students reflect on their emotions to be able to reason why they felt as they did and use this knowledge to advantage.

SKEMP'S THREE MODES OF BUILDING AND TESTING

In his book *Intelligence, Learning and Action*, Richard Skemp (1979, p. 163) made a valuable distinction between different modes of building and testing conceptual structures shown in Table 1. He speaks of building and testing a personal 'reality' as opposed to the 'actuality' of the physical world. Mode (i) relates to the individual's conception of the world we live in ('actuality'), mode (ii) to the individual's

Table 1. Skemp's Modes of Building and Testing

REALITY CONSTRUCTION	
REALITY BUILDING	**REALITY TESTING**
Mode (i) from our own encounters with actuality: *experience*	*Mode (i)* against expectation of events in actuality: *experiment*
Mode (ii) from the realities of others: *communication*	*Mode (ii)* comparison with the realities of others: *discussion.*
Mode (iii) from within, by formation of higher order concepts: by extrapolation, imagination, intuition: *creativity*	*Mode (iii)* comparison with one's own existing knowledge and beliefs: *internal consistency.*

THE DEVELOPMENT OF MATHEMATICAL THINKING

relationships with others, and Mode (iii) to the individual's relationship with mathematics itself. There is a strong relationship with the levels of *Thinking Mathematically* (convince yourself, convince a friend, convince an enemy), in terms of order of levels, but not in a one-to-one fashion. Whereas Mode (i) refers to the personal perceptions of the world based on experience and reflections on actual experiments, the act to 'convince yourself' can involve any personal ideas that the individual may bring to bear on the problem in hand. However, in both cases, the onus is on the individual to use their own resources. Meanwhile Mode (ii) involves relationships with others, which would include both friends and 'enemies', where the latter are doubters who demand a higher level of rigour. Skemp's beautiful Mode (iii) involves the relationship of the human mind and spirit with mathematics, through creativity and internal consistency.

In *Thinking Mathematically*, the role of Mode (iii) is formulated in terms of an 'internal enemy', in which the individual learns to criticise their own creative thinking to seek self-improvement and internal consistency. The full list of levels of explanation in *Thinking Mathematically* is therefore:

> Convince yourself
> Convince a friend
> Convince an enemy
> Develop an internal enemy.

Long-term this leads to the desire to think mathematically by producing arguments that may begin with personal insights, are made clearer by discussions with a friend, then with an enemy whose purpose is to challenge the ideas put forward and make the deductions more rigorous. The ultimate goal is a personal level of consistency corresponding to a Mode (iii) relationship with the coherence of mathematical ideas themselves.

MATHEMATICS AND THE EMOTIONS

Thinking Mathematically focuses on the role of the emotions in mathematics, particularly in dealing with the high of an 'Aha!' experience which should be enjoyed before subjecting the insight to further scrutiny, and being 'Stuck', requiring a positive approach to analyse what has happened and how this can help to suggest alternative approaches.

In the middle of the twentieth century, psychologists separated the cognitive and affective domains (as, for instance, Bloom's famous *Taxonomy of Educational Objectives* distinguished three distinct domains: cognitive, affective and psychomotor). Richard Skemp stood out from the crowd by relating the cognitive and affective domains in terms of his (1979) theory of goals and anti-goals. A goal is an intention that is desired. It may be a short-term simple goal, for instance, to add two numbers together, or it may be a long-term major goal, for example, to succeed in mathematics. On the other hand, an anti-goal is something that is not desired and is to be avoided. For instance, a child may wish to avoid being asked a question in class because of a fear of being made to seem foolish. In general terms a goal is something that increases the likelihood of survival, but an anti-goal is something to avoid along the way.

Children are born with a positive attitude to learning. They explore the world spontaneously, with great pleasure. But unpleasant experiences may cause them to avoid a repetition of that unpleasantness, which leads to the development of anti-goals.

In his theory of goal-oriented learning, Skemp formulated two distinct aspects of goals and anti-goals. One concerns the emotions sensed as one moves towards, or away from, a goal or anti-goal (represented by arrows in Figure 1). The other concerns an individual's overall sense of being able to achieve a goal, or avoid an anti-goal (representing by the smiling faces for a positive sense and frowning faces for a negative).

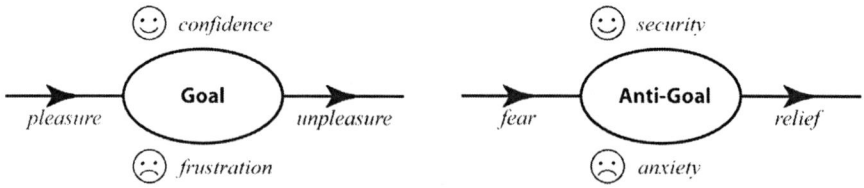

Figure 1. *Emotions associated with goals and anti-goals*

The emotions related to goals and anti-goals are very different. Believing one is able to achieve a goal is accompanied by a sense of confidence, whilst being unable to achieve a goal is accompanied by frustration. Moving towards a goal gives pleasure, whilst moving away gives unpleasure, in the sense employed earlier by Freud. It is subtly different from the more usual, but not equivalent, term 'displeasure'. Drifting away from a desired goal may not be 'unpleasant' in the sense that it is distasteful, it may simply generate a feeling that one is going on the wrong path and intimate the need to reconsider one's options.

By using Skemp's theoretical framework while working with the book *Thinking Mathematically*, I found it possible to have discussions about individuals' emotional reactions to mathematics, to recognize the different emotional signs and to use them to advantage. For instance the subtle difference between frustration and anxiety in being unable to solve a problem reveals the difference between a goal one desires positively and an anti-goal one wishes to avoid. Once the source of the problem is identified, it becomes possible to take action to move in a more appropriate direction.

PROOF ANXIETY

The one important word missing from *Thinking Mathematically* is 'proof'. In a private conversation, John told me that this was because of the reaction of students to the word in his summer schools working with Open University students. If the idea of 'proof' was mentioned, they froze. In Skemp's terminology this seems to be anxiety arising from a sense of not being able to avoid an anti-goal. Proof seems to be something that these mature students had difficulty with, and they had long since seen it as a topic that they wished to avoid. If guided towards it, they felt a sense of fear, which could only be relieved by moving away from it again.

Thinking Mathematically is designed to give positive encouragement to students through strategies that are likely to lead to the pleasure of success and build confidence in the art of problem solving as a goal to be achieved, rather than an anti-goal to be avoided. So what is it that causes proof to become an anti-goal? To gain insight into this, it is helpful look at the long-term development of mathematical thinking.

Cognitive development of mathematical thinking

In a number of recent papers (e.g. Tall, 2008), I have followed the path of development of human thinking from mental facilities *set-before* birth and the subsequent experience *met-before* in our lives that affect our current thinking as it matures. Long-term we develop through refining our knowledge structures, coming to terms with complicated situations by focusing on important elements and naming them, so that we can talk about them and build ever more sophisticated meanings. Mason's insight of a *delicate shift of attention* plays its part in switching our thinking from the global complications to the essential aspects that turn out to be important. More generally it is the *discipline of noticing* that is important to seek to focus on essential ideas and gain insight into various problematic situations.

The framework that I have developed centres on the way in which we use words and symbols to *compress* knowledge into thinkable concepts, such as compressing counting processes into the concept of number or the likenesses of triangles into the principle of congruence in Euclidean geometry. Through experience and reflection, we build thinkable concepts into knowledge structures (schemas) that enable us to recognise situations when we attempt to solve new problems. Problem Solving arises when our knowledge structures are not sufficient to recognise the precise problem, or, if we have recognised it, to have the connections immediately available to solve it. To be more effective in mathematical thinking we therefore need to be aware of how our knowledge structures operate and how they develop over time.

As a pupil of Richard Skemp, I was taken by his simple analysis of the way the human mind works through *perception, action* and *reflection*, which gives us input through perception, output through action and makes mental links between the two through reflection. Skemp took his theory forward by suggesting that the mind operated at two levels, delta-one with physical perception and action, and delta-two with mental perception and action, linked together by reflection. I reflected on this structure and came to the conclusion that the distinctions between what we *perceive* through our senses and what we *conceive* in our mind are not as clear as we might wish them to be. So, rather than a two-stage theory, I saw a developing mental structure focusing on the complementary nature of perception and action and how it shifts from physical perceptions and actions to mental structures.

Quite recently (February 2008 to be more precise) I realised, to my astonishment, that our mathematical thinking could be seen to develop from just *three* mental facilities that are set-before our birth and which come to fruition through our personal and social activities as we mature. I termed these three set-befores: *recognition, repetition* and *language.* Recognition is the human ability, which we share with many other species, of recognising similarities and differences that can be *categorised* as thinkable concepts. Repetition is the human ability, again shared with other species, of being able to learn to repeat sequences of actions in a single operation, such as see-grasp-suck, or the human operations of counting or solving linear equations. This is the basis of *procedural* knowledge. However, language enhances the set-befores of recognition and repetition. Recognition can be extended to give successive levels of thinking: forming thinkable concepts, then using those concepts as mental objects of attention to work at higher levels. Repetition can be compressed subtly through *encapsulation* of operations as thinkable concepts, denoted by symbols that can evoke either the underlying operation to perform, or the thinkable concept itself to be manipulated in its own right. These thinkable concepts that act dually, ambiguously and flexibly as process and concept are named *procepts*. As thinking processes become more sophisticated, language itself becomes increasingly powerful, leading to new formal ways of forming concepts through *definition* and mathematical *proof.*

This offers a framework for the development of mathematical knowledge structures, building on recognition, repetition and language, with compression into thinkable concepts through categorisation, encapsulation and definition, evolving through three distinct but interrelated mental worlds of mathematics that I term conceptual embodiment, proceptual symbolism and axiomatic formalism. Within the confines of this framework I usually compress the names to single words: embodiment, symbolism and formalism, while acknowledging that these terms have very different meanings in other theories.

This enables me to put the names together in new ways, such as formal embodiment, embodied symbolism, or formal symbolism. Indeed, the meanings of the two word phrases themselves depend on the direction travelled. Arithmetic arises from counting, adding, taking away, sharing as embodied operations that shift into symbolic embodiment. Representing number systems on the number line shifts back to give an embodied symbolism.

For instance, algebra builds from embodiment to symbolism through generalised arithmetic operations of combining, taking away, sharing, distributing, and so on. The reverse direction takes us from algebraic expressions and functions to graphs. These are quite different activities and, as we shall see later, there are a number of problematic aspects of these relationships.

The cognitive development of proof in the embodied world

We now turn our attention to see what the framework of embodiment, symbolism and formalism tells us about students' growing appreciation of proof.

In the embodied world of geometry, building on perception of figures and actions to make constructions gives us more specific insight into the nature of these figures. We already have the analysis of van Hiele to chart the development over the years. Give a child a plastic triangle, with equal sides and the child sees it as a whole and can touch and explore it to sense its corners, its sides and its angles. At one and the same time, it has three equal sides *and* three equal angles. From this beginning, where a figure has simultaneous properties, the child moves through successive van Hiele levels where the meanings and relationships change in conception. I choose to describe these successive levels as:

Perception: recognising shapes

Description: verbalising some of the properties

Definition: prescribing figures in terms of selected properties

Euclidean Proof: using constructs such as congruent triangles to build up a coherent theoretical framework of Euclidean geometry

Rigour: Formulating other geometric structures in terms of set-theoretic axioms.

In school mathematics, we are mainly concerned with the first four levels up to the development of Euclidean proof. My major focus of attention is the shift from Description to Definition. It seems innocuous. One simply moves from specifying certain properties of a figure to giving a more focused definition. However, cognitively, there is a *huge* shift in meaning. The plastic triangle that the child describes as being equilateral with its three equal sides and three equal angles is now defined as having three equal sides. Full stop.

The child can *see* that an equilateral triangle also has three equal angles, but now it becomes necessary to *prove* that an equilateral triangle, as defined, really does have three equal angles, *as a consequence* of having three equal sides. The method of proof is quite technical. It goes like this. First establish the meaning of congruent triangles. (Two triangles are congruent if they have three corresponding properties: three sides, two sides and included angle, two angles and corresponding side, or right-angle, hypotenuse, one side.)

Effectively the notion of congruence depends on embodied actions. If two triangles ABC, XYZ have two sides equal $AB = XY$, $AC = XZ$ and included angle equal, $\angle A = \angle X$, then pick up triangle ABC and place it on triangle XYZ with vertex A placed on X, side AB placed on XY and angle A over angle X. Then, because the angles are equal, the side AC will lie directly over XY and, because the side-lengths are equal, point C will be coincident with X and point B will be coincident with Y. It follows that all the other corresponding aspects must be equal, including all corresponding angles, all corresponding sides and even the midpoints of the respective sides, the angle bisectors, and so on.

Now take a triangle ABC with equal sides AB, BC and, by constructing the midpoint M of the base AC, form two triangles ABM and CBM. These have corresponding sides equal, $AB = CB$ (given), $AM = CM$ (by construction), BM (common), so the triangles are congruent and, in particular, $\angle A = \angle C$. Q.E.D. Apply the same argument again, and if a triangle has three equal sides, *then* it has three equal angles.

There are some who appreciate the need for proof and get great pleasure out of the beauty of many aesthetic ideas in Euclidean geometry, such as the circle theorems where two angles subtended by the same chord in a circle are equal. But the vast majority of learners have connections in their minds that *tell* them such things as the fact that an equilateral triangle has equal sides and equal angles, and so, *why do they need to 'prove' it*. The shift from description to definition and deduction is mystifying for many and forms an obstacle causing fear and anxiety. Indeed, the only way to cope with the problem is to use the met-before of repetition to learn the proofs as procedures by rote. It addresses the goal of passing examinations without attending to the goal of understanding.

The cognitive development of proof in the symbolic world

The symbolic world of arithmetic and algebra develops out of embodied actions of counting, adding, taking away, making a number of equal-sized groups, sharing, and so on. These are then symbolised and there is a shift of attention away from specific embodiments and towards the relationships between the symbols.

In the embodied world of counting, it is not initially obvious that addition is commutative. If a child is at a stage of 'count-on' then $8 + 2$ by counting on two after 8 to get 9, *10* is much easier than count-on 8 after 2 to get 3, 4, 5, 6, 7, 8, 9, *10*. The realisation that it is possible to perform the shorter count and get the same answer can be a pleasurable moment of insight.

Over time, experience shows that addition and multiplication are independent of order, and do not depend on the sequence in which the operations are performed, so that $3 + 4 + 2$ can be performed as $3 + 4$ is 7 then $7 + 2$ is 9, or as $4 + 2$ is 6 and $3 + 6$ is also 9. These are formulated as 'rules', though they are not rules that are to be imposed on numbers, but observations that have been noticed. Then there is the associative law that says that $3 \times (4 + 2)$ is the same as $3 \times 4 + 3 \times 2$, which gets more interesting in sums like $20 - 3 \times (4 - 2)$ being the same as $20 - 3 \times 4 + 3 \times 2$.

At this stage the learner has to deal with a range of principles in using the notation of arithmetic and how they operate in practice. These principles are then employed in algebra.

To 'prove' the formula for the difference between two squares, it is usual to start with $(a + b)(a - b)$ and to multiply it out using the 'distributive law' then use commutativity of multiplication to reorganise the expression and cancel ba and $-ab$ to get the final result:

$$(a + b)(a - b) = a(a - b) + b(a - b)$$
$$= a^2 - ab + ba - b^2$$
$$= a^2 - b^2$$

The problem here is to know what is 'known' and what needs to be 'proved'. The 'laws' being quoted (if they are indeed spoken explicitly) depend on experience and build on all kinds of met-befores that are implicit within the mind. While it may be appropriate in the more sophisticated axiomatic formal world to build proofs on definitions and deductions, for the teenager struggling with algebra it may cause nothing but confusion.

My own view is that the shift from embodiment to symbolism that operates in whole number arithmetic is not as evident in the shift from embodiment to algebra. For the learner who has a flexible proceptual view of symbolism, algebra may be an easy, even essentially trivial, application of generalised arithmetic. But for the learner who is already struggling with arithmetic and operates more in a time-dependent, procedural manner, it is likely to be highly complicated.

Letters may be used to represent unknown numbers in an equation such as $3x + 5 = 5x - 7$ or as units as in 120 cm = 1.2 m. The famous 'students and professors problem' relating the number of students (S) to the number of professors (P) when there are 6 students for each professor should be written as $S = 6P$ using the algebraic meaning of letters. However, it is often interpreted as $1P = 6S$ in the units sense that 1 professor corresponds to 6 students.

The met-before that every arithmetic expression, such as $3 + 2$, 3.14×4.77, or $\sqrt{2} + 1$, 'has an answer' is violated by algebraic expressions such as $3 + 2x$ that has no 'answer' unless x is known. So now the student who is bewildered by expressions that cannot be worked out is asked to manipulate them as if he or she knows what they are, when they have no meaning.

The interpretation of letters as objects that may help the student simplify $3a + 4b + 2a$ to $5a + 2b$ by thinking of a as 'apple' and b as 'banana', but it fails to give a meaning to the expression $3a - 5b$. (How can you take away 5 bananas when you only have 3 apples?)

The idea that an equation such as $5x + 1 = 3x + 5$ is a balance between 3 things and 5 on one side and 5 things and 1 on the other is seen as being widely meaningful to many students (Vlassis, 2002). Take $3x$ off both sides to get $2x + 1 = 5$, now take 1 off both sides to get $2x = 4$ and divide both sides by 2 to get the solution $x = 2$. But change the equation slightly to $3x + 5 = 5x - 7$ and suddenly it has no embodied meaning. How can you imagine a balance in which one side is $5x - 7$? How can you take 7 away from $5x$ when you don't yet know what x is?

In so many ways, the shift from embodiment to symbolic algebra is a minefield of dysfunctional met-befores for so many learners. This does not lead to the goal of making sense of algebra to develop power in formulating and solving equations. Instead, algebra becomes a topic to be avoided at all costs, an anti-goal provoking fear and a sense of anxiety as one attempts to find any method possible to avoid failure. For so many it leads to dysfunctional ways of learning procedures to cope with the difficulties: the English use of BODMAS to remember the order of precedence of operation (Brackets, Of, Division, Multiplication, Addition, Subtraction), the American acronym FOIL to multiply out pairs of terms in brackets (First, Outside, Inside, Last), operations to solve equations such as 'change sides, change signs; divide both sides by shifting the quantity to the other side and put it underneath.' For so many, algebra is an anti-goal to be avoided at all costs.

Now we are beginning to build up a picture of what may be happening in school as children learn arithmetic, then algebra. For so many, the initial embodiments of putting together and sharing have a meaning in the actual world in which they live. But the many successive compressions in meaning from operation to flexible procept work for some but impose increasing pressures on others. Eddie Gray and I called this 'the proceptual divide' in which the flexible thinkers have a built-in engine to derive new facts from old based on their rich knowledge of relationships between numbers, while others see increasing complication in all the detail and fall back on attempts to learn procedures by rote to cope with the pressures of testing.

Learning procedures by rote can be supporting in being able to perform routine calculations but procedural learning alone makes it more difficult to imagine flexible relationships between compressed concepts that are required in more sophisticated problem solving. As mathematics becomes more complicated for those who lack the rich flexible meanings, mathematics itself becomes an anti-goal to be avoided, creating a sense of anxiety and fear. More generally, mathematical proof, which requires a coherent grasp of ideas and how they are related, becomes problematic, both in geometry and in algebra.

GENERATING CONFIDENCE THROUGH *THINKING MATHEMATICALLY*

Given the relationship between cognitive success and emotional reactions, it becomes likely that one might attempt to improve students' abilities to think mathematically through organising situations in which they may experience success. Having experienced the good feelings generated in an open-ended problem-solving course myself, I was fortunate to be joined by Yudariah binte Mohammad Yusof, a university teacher from Malaysia who was concerned by the concentration on procedural learning in her students and the lack of a problem-solving ethic, other than that of becoming highly proficient at solving specific problems that would feature on the university examinations.

She took part in the Problem Solving course at Warwick University and trialled a questionnaire investigating student attitudes towards various aspects of mathematics and problem-solving. She then returned to Malaysia to teach the course and to research its effect on the students. (The details are given in Yusof & Tall 1996.) Half way through the course she telephoned me to express concern that her students continued to ask her what she wanted them to do, so that they could do well on the course. All I could say to her was that she should maintain the objective that the students needed to take control of their own working using the framework of *Thinking Mathematically*.

By the end of the course attitudes had changed dramatically. To identify what was meant by a 'desirable change', she asked the students' lecturers to fill in the questionnaires twice, once to indicate what they *expected* the students to say, once to say what they *preferred* the students to say. The direction of change from expected to preferred was taken to be a 'positive' change. In general *all* the changes in students' attitudes during the problem-solving course were positive, but when they returned to their normal mathematics lectures and were asked again six months later, the changes generally went back in the opposite direction. In other words, the problem-solving course took the students' attitudes in the direction desired by the staff, but when the staff themselves did the teaching, the attitudes of the students changed to the opposite direction.

My experience to date, through the work of research students carrying out studies in other countries and through my own links with communities around the world, is consistent with the global concern about the learning of mathematics. Some societies try to encourage meaningful learning through problem-solving, some teach by rote to encourage proficiency with many variations in between. Everywhere the pressure to compete and succeed in tests is driving the policies of governments. Surely our job as mathematics educators is not just to increase percentages passing examinations but a wider and deeper concern to understand the nature of mathematical thinking, to identify precisely why it is so difficult for many and how it can be improved for each individual.

REFLECTIONS

The analysis given here shows the power of *Thinking Mathematically* to improve students' attitudes and improve students' self-confidence and pleasure in doing mathematics and thinking for themselves. However, this occurs in a context in which so many older students have anxieties in dealing with the most central of all mathematical concepts, the notion of *proof*. The analysis given here in terms of the development of mathematical thinking through increasingly sophisticated embodiment and symbolism reveals transitions that are required to make sense of increasingly sophisticated mathematical thinking. The apparently innocuous shift from description to definition in geometry violates earlier beliefs in the properties of figures that are 'known' as part of a global perception but now must be 'proved' from the selected definitional properties. The shift from arithmetic to algebra involves a range of met-befores where established beliefs need to be changed to make sense of the new ideas.

John Mason has led a personal crusade for everyone to think about mathematics in new ways and his methods have yielded success. Clearly the way forward is to increase students' confidence by giving them genuine experiences of successful thinking, for only then will they face new problems as a challenge rather than a source of anxiety and fear. There is still much to be done by future generations to extend the pathways already trodden.

REFERENCES

Bloom, B. S. (1956). *Taxonomy of educational objectives, handbook 1: The cognitive domain*. New York: David McKay.
Mason, J., Burton, L., & Stacey, K. (1982). *Thinking mathematically*. Wokingham, UK: Addison Wesley.
Skemp, R. R. (1979). *Intelligence, learning and action*. London: Wiley.
Tall, D. O. (2008). The transition to formal thinking in mathematics. To appear in *Mathematics Education Research Journal*. Retrieved from http://www.warwick.ac.uk/staff/David.Tall/pdfs/dot2008e-merj-3worlds.pdf
Vlassis, J. (2002). The balance model: Hindrance or support for the solving of linear equations with one unknown. *Educational Studies in Mathematics*, 49(3), 341–359.
Yusof, Y. B. M., & Tall, D. O. (1996). Conceptual and procedural approaches to problem solving. In L. Puig & A. Gutierrez (Eds.), *Proceedings of 20th meeting of international group for the psychology of mathematics education* (Vol. 4, pp. 3–10). Valencia, ES.

David Tall
University of Warwick (United Kingdom)

DEREK HOLTON, MICHAEL THOMAS AND
ANTHONY HARRADINE

3. THE EXCIRCLE PROBLEM

A Case Study in How Mathematics Develops

INTRODUCTION

A large, highly polished and expertly cut diamond sits proudly in a secure and beautifully lit showcase in a London museum. It is admired by the many who view it. Few who view it, however, know much about the origins of this gem, the mine from which it came, what it looked like before it was cut and polished, the messy work of the many contributors who helped make it the beautiful object that it is today. Even fewer can lay claim to having been one of those contributors.

A similar comment can be made about mathematics. There is this beautiful object and we are shown it in all its glory. However, we are seldom shown the sometimes messy way that it got to its present state. As Thomas and Holton (2003, p. 347) note, "In research, everything is not as neat and sewn up as the mathematics that is often presented in lectures and in textbooks. First, mathematical research is about solving problems." Sadly it is true that the origins of mathematics, and the importance of solving problems, remain a mystery to many students.

In this chapter we try to give some idea of the creative process of mathematics by describing and annotating an expedition into the mine of mathematics. This process of solving problems has been very well presented by Mason, Burton and Stacey (1982), who describe three key stages in the process, namely Entry, Attack and Review, each of which folds back to the others. Figure 1 shows how the major activities of specializing (considering a few special cases), and generalizing (writing about a wide class of examples) fit into this outline.

This process is not dissimilar to the research process outlined by Thomas and Holton (2003) as comprising the following steps: problem identification; experimentation; pattern recognition; evidence-based conjecture; testing of conjecture through examples; refinement of conjecture; statement of result; and proof of result.

In what follows, we attempt to solve a number of problems, and try to relate this to the creative process of problem solving in mathematics mentioned above, and the steps involved in it.

We illustrate examples of 'mathematics *through* problem solving' or 'thinking mathematically' (Mason, Burton & Stacey, 1982). In a related discussion Schroeder and Lester (1989) talk of three types of teaching. First they describe the traditional

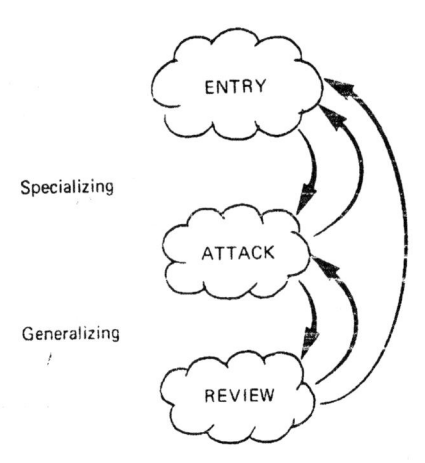

Figure 1. The problem solving process (from Mason, Burton and Stacey, 1982, p. 28).

method that is 'mathematics *for* problem solving'. The idea here is that mathematical skills are taught with a view to the students being able to use them to solve problems. Textbooks are generally designed this way with exercises sprinkled through the text to give students practice in the skills that have been demonstrated. Then, at the end of a section or chapter, problems are posed that require a little more than the use of a single skill on a standard question.

The second type of teaching is 'mathematics *about* problem solving'. The idea here is that mathematics is used to teach students how to solve problems. In this method, teachers emphasize such things as heuristics and underline the scaffolding that they are using.

Finally we have 'mathematics *through* problem solving'. This turns the first type around and uses problems to teach mathematical skills and results, as encouraged by Mason, Burton and Stacey (1982). This is what we try to do here by presenting a problem that hints at the existence of a gem but gives no ideas about where or how to dig for that gem. We then engage in a process, part of which involves gathering data that can be efficiently collected using computer software, either housed in a COT[1] calculator or on a computer. We show how the data leads to conjectures that might be proved or counter-examples found. In the latter case we move on hopefully to better conjectures and then theorems. Along the way we find some new and useful results.

We also give an example of a proof that might be found in a textbook. This compares with earlier proofs in that it is unmotivated and by itself gives no idea of the creative process behind it.

Our aim in this chapter, then, is to provide a sequence of events that we hope will enable a student to see how mathematicians might approach a problem and how they might enjoy this journey to a solution.

THE STARTING PROBLEM

Draw a right-angled triangle and extend the two non-hypotenuse sides. Now draw the circle that is tangential to the two extended sides and to the hypotenuse (see Figure 2). This circle is called the excircle.

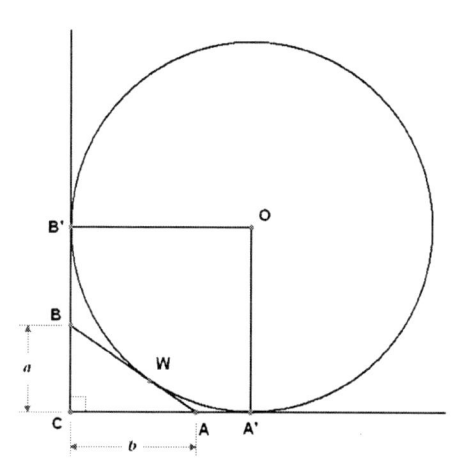

Figure 2. The circle and its tangents.

If the length of the two non-hypotenuse sides of the triangle are a and b respectively, what is the radius of the excircle?

This is, of course, an old problem and 'everyone' knows the answer. Some people even know how to prove it. But if you are an educator your students are unlikely to be in either of those categories and so we treat this as a voyage of discovery for them. We hope you find that you are able to use the ideas that follow in your classroom or lecture theatre.

It seems the natural way to begin is to study a number of specific cases of the general construction. What follows is the very kind of experimentation that some geometric software was designed to do. It's actually not clear how you could do what we are going to do without the software. We suppose a class of students could team up and draw, with ruler and compass, one case each. But that process is not easy and you could not guarantee the accuracy of the answers found by each student.

At this point we are going to assume that you have some computer software at hand that will enable your students to draw the diagram above and to gain accurate measurements of the lengths of various line segments. With appropriate software, individual students can each produce many cases with great accuracy.

Figure 3 shows the construction built in the Geometry application on the COT calculator called the ClassPad. The constraint nature of the ClassPad's geometry makes the construction process here very simple.

So what cases do we suggest? Well, it's probably best to take 'simple' right-angled triangles, where simplicity here means whole numbers for the sides of the right triangle. Such creatures are commonly called Pythagorean triples. The reason for using these is that it will make any patterns that we might get for the radius, r, of the constructed excircles, easier to identify.

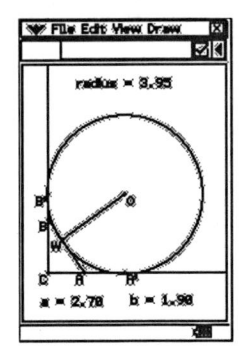

Figure 3. The construction and measurements as drawn using the ClassPad.

The 16 primitive[2] Pythagorean triples, PPTs, that are shown in Wikipedia (http://en.wikipedia.org/wiki/Pythagorean_triple) should be enough to experiment with right now.

Excircle Experimentation

In our excircle construction we can set a and b (which will automatically determine c's value since our construction demands a right angled triangle) equal to the first two numbers in your favourite PPT, and measure the radius r of the excircle as seen in Figure 4.

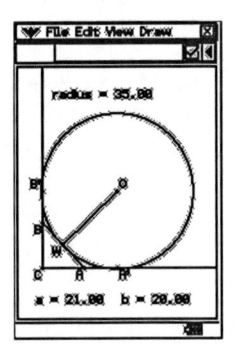

Figure 4. The PPT (20,21,29) giving an excircle radius of 35 units.

This will give us something like the Table 1 below. You might ask your students how they want to display their data and see what form they think is best for them.

THE EXCIRCLE PROBLEM

Table 1. Looking for a link between the sides of the triangle and the radius of the excircle.

a	3	5	7	8	9	11	12	13					
b	4	12	24	15	40	60	35	84					
r	6	15	28	20	45	66	42	91					

Looking for Patterns

What can we see here? Once more, in the words of Mason (1992, 1995) it depends on whether we are *looking at* the columns of numbers or *looking through* them; whether the focus of our attention is surface or deep. You will need to discuss this with your class to see what they come up with, and as you do so it is important to keep in mind at least these two things suggested by John Mason. First, that

> ... the style and format of the questions used by lecturers and tutors profoundly influence students' conceptions of what mathematics is about and how it is conducted. By looking at reasons for asking questions, and becoming aware of different types of questions which mathematicians typically ask themselves, we can enrich students' experience of mathematics. (Mason, 2000, p. 97)

Second, in spite of our best efforts there can be a communication failure due to a mismatch of attention, the possible causes of which were outlined by Mason (2008, p. 9):

if some learners are attending holistically when the teacher is discerning specific details;

if some learners are discerning details when the teacher is talking about relationships amongst details;

if some learners are recognising relationships amongst discerned elements when the teacher is talking about properties of objects in general;

or if some learners are thinking about properties when the teacher is deducing from properties;

Let us do some thinking that may or may not mirror that of a student. Experience tells us that often there will be a number of different thought paths occurring within a class of students, and a key notion in the solving process is that of what students (and the teacher) are attending to (Mason, 1989).

Hmm! 6 is twice 3; 15 is 3 times 5; 28 is 4 times 7. But 20 isn't 5 times 8 (but it is $\frac{5 \times 8}{2}$!). At this point we are beginning to try and *articulate* (Mason, 1989) the generality that we may have noticed.

Actually the pattern seems to work for odd numbers though. $45 = 5 \times 9$; $66 = 6 \times 11$; $91 = 7 \times 13$. Where are these numbers 2, 3, 4, 5, 6, 7 coming from? Could it be that for a odd,

35

$$r = \frac{a+1}{2} \times a \ ?$$

Here we are articulating the generality using "increasingly tight and economically succinct expressions" (Mason, 1989, p. 3), namely algebra, in order to get a sense of what we have. Does that fit all odd values of a? Can we get another pattern for even values of a? Can we prove that these patterns are correct?

Let's put down a conjecture (a guess).

Conjecture 1: If a is odd, then $r = \frac{a+1}{2} \times a$.

Ah! There's the rub. Mathematicians have to *prove* things. It's part of the mathematicians' creed. But it can be a nuisance. It's not always an easy thing to do. How could we prove Conjecture 1 or any of the other conjectures that your students are coming up with?

Well unfortunately the conjecture is false. Your students should be able to find (they may even have found), a counter-example – an example that is counter to the conjecture. It is also possible that you might have felt uncomfortable with this conjecture since it did not include the value of b. It seems unlikely that R is unaffected by the value of b.

There are a number of interesting relationships that can be guessed from a table.

So where do we go now? Well maybe we should look at Table 1 again. Unfortunately we didn't include c in that table. The reason that we left it out is that c depends on a and b so we guessed that it wasn't going to be worth putting c in. But maybe we should do that now (Table 2).

Table 2. Looking for a link between all of the sides of the triangle and the radius of the circle

A	3	5	7	8	9	11	12	13				
B	4	12	24	15	40	60	35	84				
C	5	13	25	17	41	61	37	85				
R	6	15	28	20	45	66	42	91				

Does that help? In all cases, and not just for odd a, there's a guessable link between a, b, c and r. Can we articulate it?

Conjecture 2: $r = \frac{a+b+c}{2}$.

So, can we prove this or can we find a counter-example? Maybe it's too pretty a possible result to be wrong. Mathematicians always head for the nicest possible links first. They may not always be right but neatness guides their guesses.

THE EXCIRCLE PROBLEM

Looking for a proof

So where can we find *r* in the diagram and can that length be shown to be what we hope it is? If you look at Figure 5, you'll notice that the two line segments marked *r*, are two adjacent sides of a square. The other sides are *a* plus a bit (BB') and *b* plus a bit (AA'). Can we say anything about those bits? Ah, let's call them *p* and *q*. So we have $r = a + p$ and $r = b + q$.

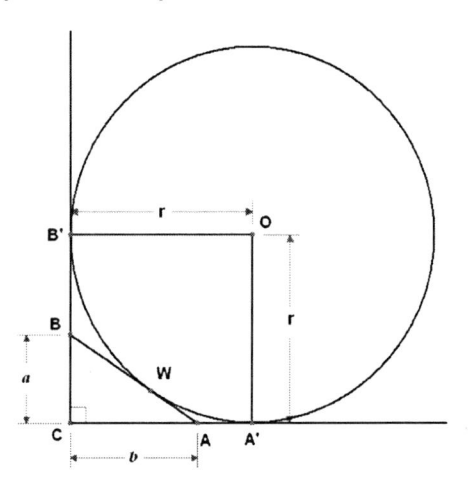

Figure 5. A square, OB'CA' can be identified in our construction.

But what are *p* and *q*? Can we find them anywhere else on the diagram? Fortunately yes. BB' and BW are segments of the two tangents to the circle from B.

So p = BW = BB'.

In a similar way, q = AW = AA'.

But what good is that? We don't know either BW or AW so we haven't got any further! Wait though. We know BW + AW, that's the hypotenuse of the triangle. So BW + AW = *c*. So now we have three equations.

$$r = a + p \text{ and } r = b + q \text{ and } p + q = c.$$

So we get $2r = a + b + p + q = a + b + c$ and of course this gives us

$$r = \frac{a+b+c}{2}.$$

And now Conjecture 2 is proven – a triumph for beauty; well mathematical beauty anyway. And that wasn't so hard was it? In fact we think that making such an argument is well in the grasp of school-aged students and that many 13-year-old students would be able to see and understand this proof.

What next?

When you think about it we concentrated solely on the one excircle. What about the excircles on the non-hypotenuse sides? Can we find a nice expression for them? Of course we haven't even thought about the incircle. Wouldn't you like to know what the radius of the incircle is? (Or maybe you'd just like to know what an incircle is!) Can we get that in terms of a, b, and c?

But why stick with right angled triangles? What if we are dealing with equilateral triangles or isosceles triangles or any old triangle you might like to name? Can we find the radius of the in- and excircles in all of these cases? Would this give us a generalization?

What did we do in the proof that actually required the right angle? The point of asking this is to see if the excircle on the side AB is always going to have radius

$$r = \frac{a+b+c}{2}.$$

Unfortunately we needed the right angle to be able to say that there was a square with side length r. But the algebra we did along the sides didn't use the angle at C at all. No matter what the angle at C, the excircle on AB would have divided the length of AB into two parts of length:

$$p = \frac{-a+b+c}{2} \quad \text{and} \quad q = \frac{a-b+c}{2}.$$

Perhaps that might come in handy later.

THE OTHER EXCIRCLES

So how do we get started here? Well, what worked for us before? We experimented to get a conjecture and then we looked for a proof. Let's try it again.

Suppose that we try to get the radius, r_a, of the excircle on the side of length a. We need to construct and work on the diagram shown below (Figure 6). The table that we need to fill here (Table 3) will be almost the same as Table 2. And we get the value of r_a by measuring it in our COT calculator.

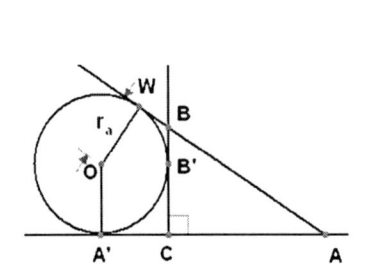

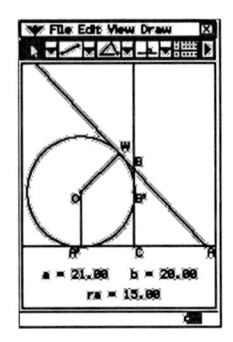

Figure 6. The excircle drawn on side BC and as seen on our COT calculator.

THE EXCIRCLE PROBLEM

A series of measurements provides us with Table 3.

Table 3. Looking for the radius, r_a, of the excircle on side a

a	3	5	7	8	9	11	12	13				
b	4	12	24	15	40	60	35	84				
c	5	13	25	17	41	61	37	85				
r_a	2	3	4	5	5	6	7	7				

Recalling our result for r, it is not too hard to make the following conjecture.

Conjecture 3: $r_a = \dfrac{a-b+c}{2}$.

The expression for r_a looks suspiciously like something we've met before. But what?

To get a conjecture for r_b, the radius of the excircle on side b, we can work two ways. We can construct the obvious table. That should serve us as well as we have been served by Tables 3 and 4. But we could also think a bit. There ought to be some symmetry between the a and b situations. So perhaps the next conjecture looks like this.

Conjecture 4: $r_b = \dfrac{-a+b+c}{2}$.

Now it's proof time again.

Proof Time

We could try using the same ideas as we did in the proof of Conjecture 2 but how can we get more than one expression involving the segments equivalent of p (B′B) and q (CB′)? That doesn't seem to be easy. But drawing a diagram like that in Figure 6 (Figure 7) looks like a good first strategy. Then maybe with a bit of thought …… And maybe by adding the odd line or two …… Does any of that help?

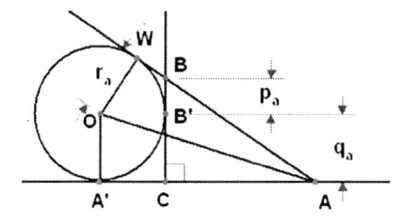

Figure 7. Figure 6, but with the inclusion of the line joining A and O.

One of the things that we notice is that we have a square in this diagram, OB´CA´. So $r_a = q_a$. Can we get something on the diagram that involves p_a? It's hard to see what we might do with p_a on the side BC but BW is also equal to p_a. Is that any help?

The line OA is interesting. It splits the quadrilateral WAA´O into two congruent triangles. How can we use that information? Ah!

$$\frac{OW}{WA} = \frac{OA'}{A'A}.$$

So what? Do we know any of those lengths? Of course $OW = OA' = r_a$, but is that any help? Do we know, or can we find, WA and A´A? Yes, A´A is just $r_a + b$ (A´C + CA). But what about WA? That has to be $p_a + c$ but is that any help? Let's put all that in the last equation.

$$\frac{r_a}{p_a + c} = \frac{r_a}{r_a + b}$$

so $p_a + c = r_a + b$. That gives us $r_a = p_a + c - b$.

That's not looking too good. First of all we want a 2 in the denominator and second we don't know what p_a is.

Oh! But we do know that $p_a + q_a = a$, so $r_a = a - q_a + c - b$. Is this just going round in circles? Yes, until we remember that $q_a = r_a$ and then it's all downhill. Conjecture 3 is true all right. And surely we can prove that Conjecture 4 is true by the same method?

It's worth writing down a Theorem to cover what we have done so far so that we can be sure of what we have actually achieved.

Theorem 1: Let a, b and c be the lengths of the sides of a right-angled triangle, where c is the length of the hypotenuse. Then

– the radius of the excircle on the hypotenuse is $\frac{a+b+c}{2}$;

– the radius of the excircle on the side of length a is $\frac{a-b+c}{2}$;

– the radius of the excircle on the side of length b is $\frac{-a+b+c}{2}$; and

– the radius of the incircle is $\frac{a+b-c}{2}$.

Of course we haven't actually proved (iv), or even conjectured it for that matter, but we're sure that you have either proved it already or can do so with the hint that's coming a little later.

What's left?

In this section we generalize Theorem 1 to any triangle whose sides are of length a, b, and c. We have avoided the 'mathematics *through* problem solving' approach here to save space and because we feel that you know by now how it might be done. But we also want to emphasize that what you usually see in a textbook is what we have put here. In this presentation there is no hint of how the result might ever have been achieved in the first place and so no hint to students how new results in mathematics might be produced.

We do suggest, however, that you might try to find the radius of the ex- and incircle for special cases first. Equilateral triangles and isosceles triangles might be a good place to start.

Theorem 2: Let ABC be any triangle with side lengths $AB = c$, $BC = a$, and $CA = b$. Then
- the radius of the excircle on side AB is

$$\frac{a+b+c}{2} \sqrt{\frac{2ab+c^2-a^2-b^2}{2ab+a^2+b^2-c^2}};$$

- the radius of the excircle on side BC is

$$\frac{a+b+c}{2} \sqrt{\frac{2ab+c^2-a^2-b^2}{2ab+a^2+b^2-c^2}};$$

- the radius of the excircle on side CA is

$$\frac{a+b+c}{2} \sqrt{\frac{2ab+c^2-a^2-b^2}{2ab+a^2+b^2-c^2}}; \text{ and}$$

- the radius of the incircle is

$$\frac{a+b-c}{2} \sqrt{\frac{2ab+c^2-a^2-b^2}{2ab+a^2+b^2-c^2}}.$$

Proof: (i) As before, $a + p = b + q = u$ and $p + q = c$ (see Figure 8). Hence $u = \frac{a+b+c}{2}$.

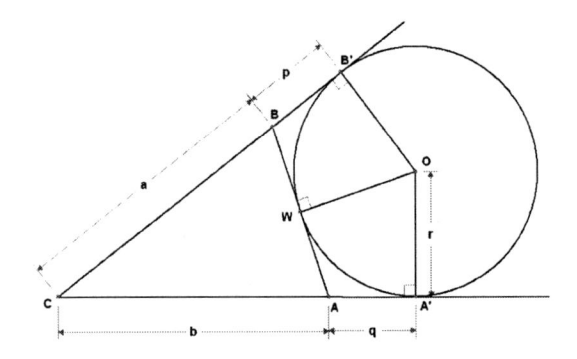

Figure 8. The general triangle ABC and its excircle on side AB.

From triangle COA′, $\tan(\frac{C}{2}) = \frac{r}{u} = t$. But $\frac{1-t^2}{1+t^2} = \cos C = \frac{a^2+b^2-c^2}{2ab}$.

Hence $t^2 = \frac{2ab+c^2-a^2-b^2}{2ab+a^2+b^2-c^2}$. The result follows from there.

Proof of (ii). We can go through the steps of part (i) but because of the symmetry of the situation we only need to interchange a and c and the result follows.

Proof of (iii). Just interchange b and c.

Proof of (iv). There's a little more to this unfortunately; but not too much more.

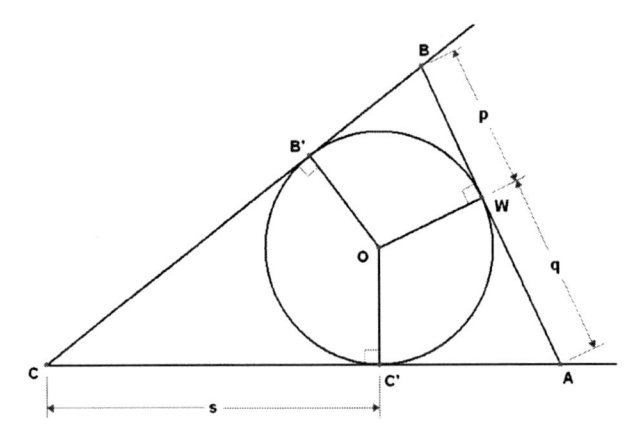

Figure 9. The general triangle ABC and its incircle.

Here $a = s + p$, $b = s + q$ and $c = p + q$ (see Figure 9). So $s = \frac{a+b-c}{2}$.

Now $r = s \tan \frac{C}{2}$. Since $\tan \frac{C}{2}$ is the same as in (i) the result follows.

THE EXCIRCLE PROBLEM

This is how the result might appear in a textbook. It provides the bare statement of the proof without any recognition of all the hard work that has gone before. Something akin to that glamour-puss diamond in the museum.

Two worries

There are two things that are still worrying us though. The first of these is the general incircle of Theorem 3. You see the expression for the length of the incircle radius should be the same whether or not we approached it from the vertex C or from A or from B. The expression should be symmetric: it should be possible to change c to a and a to b and b to c and get the same result but it doesn't look as if that is the case. At a first glance we seem to have two different expressions: the first one that we derived, namely,

$$\frac{a+b-c}{2}\sqrt{\frac{2ab+c^2-a^2-b^2}{2ab+a^2+b^2-c^2}}$$

and the one obtained from that by changing c to a to b to c, namely

$$\frac{b+c-a}{2}\sqrt{\frac{2bc+a^2-b^2-c^2}{2bc+b^2+c^2-a^2}}.$$

But those two expressions look completely different. What have we done wrong? Let's play around a bit with the first expression.

$$\frac{a+b-c}{2}\sqrt{\frac{2ab+c^2-a^2-b^2}{2ab+a^2+b^2-c^2}} = \frac{a+b-c}{2}\sqrt{\frac{c^2-(a-b)^2}{(a+b)^2-c^2}}$$

$$\left(\frac{a+b-c}{2}\right)\sqrt{\frac{(c-(a-b))(c+(a-b))}{((a+b)-c)((a+b)+c)}} = \frac{1}{2}\sqrt{\frac{(a+b-c)(c-(a-b))(c+(a-b))}{((a+b)+c)}}$$

$$= \frac{1}{2}\sqrt{\frac{(a+b-c)(c-a+b)(c+a-b)}{(a+b+c)}}$$

$$= \frac{1}{2}\sqrt{\frac{(a+b-c)(-a+b+c)(a-b+c)}{(a+b+c)}}$$

$$= \frac{1}{2}\sqrt{\frac{(a+b-c)(a-b+c)(-a+b+c)}{(a+b+c)}}.$$

Now we can see that the final expression definitely is symmetric. If we change c to a to b to c the expression for the length of the incircle radius is unchanged.

It's worth pointing out at this stage that there are at least one other way of tackling this general incircle question. One way is to use area. Look at the diagram below.

43

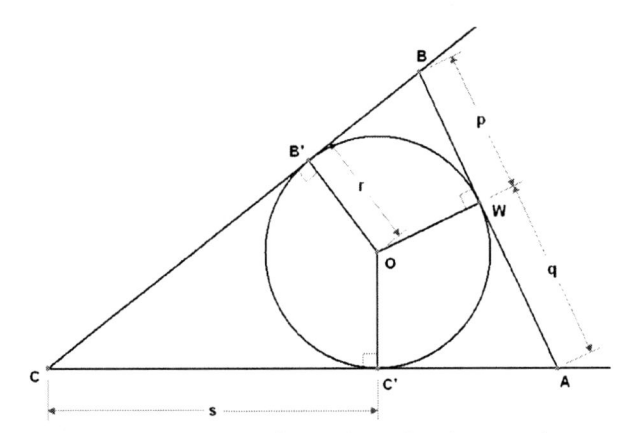

Figure 10. The general triangle ABC and its incircle.

The area of the triangle is the sum of the three kite shapes (see Figure 10), which is *pr* + *qr* + *sr*. And by Heron's formula, the area of the triangle is $\sqrt{s(s-a)(s-b)(s-c)}$, where *s* is half of the perimeter of the triangle. So the radius of the incircle is

$$\frac{2\sqrt{s(s-a)(s-b)(s-c)}}{(a+b+c)}.$$

This expression is symmetric and by replacing *s* with $\frac{a+b+c}{2}$ gives the symmetric expression we have just derived above.

A point that we want to make in passing here is that things can often be proved in more than one way. We also want to add that here we have had a case of symmetry in an algebraic expression, so symmetry isn't just confined to geometric figures, though it is motivated by it to some extent here. There are times when checking for the obvious symmetry in an algebraic expression will show up an error; fortunately for us that didn't happen this time.

The second thing that is worrying us is why the original excircle radius in Table 1 always seemed to be a product. Is this actually the case? Do we always get a product if we use Pythagorean triples as the sides of our right-angled triangle? To explain this product we need to recall that the general form for PPTs (see http://en.wikipedia.org/wiki/Pythagorean_triple) is ($2dk$, $d^2 - k^2$, $d^2 + k^2$), where *d* and *k* are coprime. Now we know the general form for PPTs this product can be easily explained. Let's do the case for the radius of the excircle on the hypotenuse. Here $r = \frac{a+b+c}{2}$. This can now be written as

$$r = \frac{2dk + d^2 - k^2 + d^2 + k^2}{2} = d(d+k)$$

Not only is this clearly a product but it also tells us what the terms in the product are. Those mysterious factors aren't a mystery any longer.

USE OF TECHNOLOGY

Part of our work involved gathering data, and in our case that data was efficiently collected using a COT calculator, a more than useful tool in the problem-solving process. However, as with any tool, one needs to give careful thought to how it is employed. Based on the ideas of Rabardel (Rabardel, 1995; Vérillon & Rabardel, 1995), Artigue (1997) and Guin & Trouche (1999) have made clear the distinction between *tool* and *instrument* use in mathematics. The process of instrumental genesis, comprising instrumentation and instrumentalisation, can turn out to be unexpectedly complex (Artigue, 2002). It involves the subject adapting her/himself to the tool, as well as adaptation of the tool to a particular task, deciding what it might be useful for, how it might be applied, and development of the skills needed to use it for the task. Hence, for a successful outcome the teacher needs to pay careful attention to the kind of tasks they use in order to help students simultaneously develop calculator and by-hand techniques, theory (Kieran & Drijvers, 2006), and to the *techniques,* or mathematical activity between tasks and theories, that the students will employ (Lagrange, 2000). That was also the case with our work, where we employed the calculator in the problem-solving process with a constant eye on the mathematics.

The process of instrumental genesis of technological tools such as COT calculators places some load on teachers in terms of the decision process that they engage in when deciding how and when to use them in their teaching. One of the factors influencing these decisions has been described by Thomas (Thomas & Hong, 2005a; Thomas & Chinnappan, 2008) as teachers' *pedagogical technology knowledge* (PTK). This may prove to be a useful way to think about what we, as teachers, need to know in order to teach well with technology. PTK includes not simply being a proficient user of the technology, but more importantly, understanding the principles and techniques required to teach *mathematics* through the technology. In the context of the current discussion this means appreciating the relationship between problem solving and the COT calculator. This may necessitate a change of mindset on our part, a shift of focus to a broader perspective of the implications of the technology for the learning of the mathematics. Part of the process of developing PTK involves us in the transformation of the technology into an instrument, and differentiation of qualitatively diverse ways of employing it in teaching mathematics, particularly as a means of building conceptual knowledge (Thomas & Hong, 2005b).

FINAL REFLECTIONS

Here we want to review again as per the problem solving process of Mason et al. (1982), or to look back in a Pólya-like Step 4 way, to see what we have done, what we haven't done, what we might have done, and how we did it.

First of all we managed to not only find the radii of the ex- and incircles of any right angled triangle in terms of its sides but we also did the same thing for **any** triangle. And not only did we find all of these things but we proved that they were correct too. So we actually achieved quite a lot. We established a lot of mathematical results that we didn't know before and that can be used in other situations.

It's hard to see where else to go with this problem. We seem to have done everything there is to do. Maybe there is something similar that we could tackle in three-dimensions but we can't see what that might be right now. We leave that up to you to explore.

But we can profitably think through what broad steps we have taken here. The diagram in Figure 11 gives some idea of how these steps are related. First of all there was a **problem**: *What is the radius of the excircle on the hypotenuse of a right-angled triangle?* To try to get a handle on that problem, we **experimented**, we used our COT calculator to generate some data on the basis of which to make a guess, a **conjecture**.

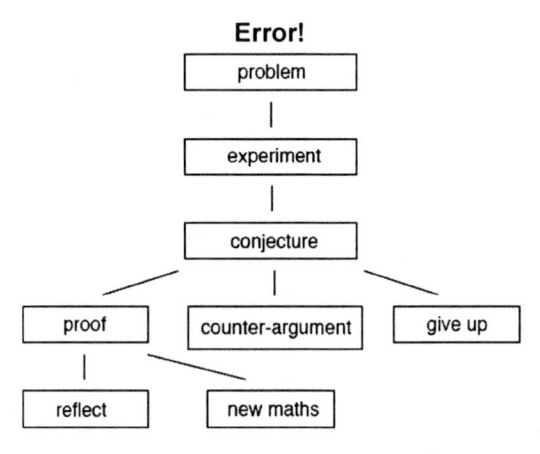

Error!

```
        problem
           |
        experiment
           |
        conjecture
       /    |    \
   proof  counter-argument  give up
     |        \
   reflect   new maths
```

Figure 11. The broad steps of the creative processes.

If our conjecture and we are good enough, then we might be able to **prove** it. If not, we may be able to find a **counter-argument** to show that the conjecture is false. In this case we might be able to find a better conjecture that we can prove. If all that fails and we have no more brilliant ideas, then we may have to **give up**. The process we have tried to exemplify with our problem solving approach above is related to what Mason and Spence (1999) describe as *knowing-to* rather than the three aspects of *knowing-about*: *knowing-that* (factual), *knowing-how* (techniques and skills), or *knowing-why* (ability to restructure actions). This *knowing-to* is a knowing-in-the-moment, an ability to notice, to mark and respond to, and to reflect upon, actions and their consequences. Developing this knowledge during problem solving involves our ability to structure our attention in the moment, to recognize what we are aware of.

In the case where we find a proof, then that proof may have led to some **new mathematics**, such as the radii of the various circles. This will go into the collection of known results ready to be used at some later point elsewhere.

About this point in proceedings it is time to **reflect** on what has happened. Have we solved the original problem completely? Are there still some open questions? Is everything OK? Do we need to check our results to see if we have made any errors? Can we generalize or extend what we have done?

It's worth noting at this point that solving problems is never as straightforward as perhaps we have made out. Although we have put single lines joining one box to the next, in practice we have to move, almost at random over the boxes. For example, in trying to find a counter-argument we might see how to prove our conjecture, reflection may lead to another conjecture, and so on.

All of these steps are steps that mathematicians will take in doing original research. About the only difference here between what we have done and what mathematicians do, is that the problems we worked on have long since been solved. However, the unique nature of the mathematical mine ensures that the process we have been through was not destroyed for us (or the students we might take there). Such an expedition is no less enjoyable and no less valuable to a student who is learning the creative process (or us in fact) than it was for those who first dug in that part of the mine. People did not stop climbing Everest after Hillary and Tensing first scaled its peak.

NOTES

[1] We use COT (Collection Of Technologies) instead of the more common CAS because most COT calculators are able to do more than algebraic manipulation. As we shall see, here we require the technology to include the manipulation of geometric shapes.

[2] Primitive ones have no factors in common between the three values.

REFERENCES

Artigue, M. (1997). Le logiciel 'DERIVE' comme revelateur de phénomènes didactiques liés à l'utilisation d'environnements informatiques pour l'apprentissage [The symbolic manipulator Derive as a revealer of didactic phenomena in the use of calculating environments for learning]. *Educational Studies in Mathematics, 33*(2), 133–169.

Artigue, M. (2002). Learning mathematics in a CAS environment: The genesis of a reflection about instrumentation and the dialectics between technical and conceptual work. *International Journal of Computers for Mathematical Learning, 7*(3), 245–274.

Guin, D., & Trouche, L. (1999). The complex process of converting tools into mathematical instruments: the case of calculators. *International Journal of Computers for Mathematical Learning, 4*(3), 195–227.

Kieran, C., & Drijvers, P. (2006). The co-emergence of machine techniques, paper-and-pencil techniques, and theoretical reflection: A study of CAS use in secondary school algebra. *International Journal of Computers for Mathematical Learning, 11*(2), 205–263.

Lagrange, J.-B. (2000). L'intégration d'instruments informatiques dans l'enseignement: une approche par les techniques [The integration of calculators into teaching: An approach using techniques]. *Educational Studies in Mathematics, 43*(1), 1–30.

Mason, J. (1989). Mathematical abstraction as the result of a delicate shift of attention. *For the Learning of Mathematics, 9*(2), 2–8.

Mason, J. (1992). Doing and construing mathematics in screenspace. In B. Southwell, B. Perry, & K. Owens (Eds.), *Space the first and final frontier, Proceedings of the 15th Annual Conference of the Mathematics Education Research Group of Australasia* (pp. 1–17). Sydney, AU.

Mason, J. (1995). Less may be more on a screen. In L. Burton & B. Jaworski (Eds.), *Technology in mathematics teaching: A bridge between teaching and learning* (pp. 119–134). London: Chartwell-Bratt.

Mason, J. (2000). Asking mathematical questions mathematically. *International Journal of Mathematical Education in Science and Technology, 31*(1), 97–111.

Mason, J. (2008). Doing ≠ construing and doing + discussing ≠ learning: The importance of the structure of attention. *Proceedings of ICME-10, Copenhagen* (CD version of proceedings). Retrieved from http://www.icme10.dk/proceedings/pages/regular_pdf/RL_John_Mason.pdf

Mason, J., Burton, L., & Stacey, K. (1982). *Thinking mathematically*. Wokingham, UK: Addison-Wesley.

Mason, J., & Spence, M. (1999). Beyond mere knowledge of mathematics: The importance of knowing-to act in the moment. *Educational Studies in Mathematics, 38*, 135–161.

Rabardel, P. (1995). *Les hommes et les technologies, approche cognitive des instruments contemporains*, Paris: Armand Colin.

Schroeder, T. L., & Lester, F. K., Jr. (1989). Developing understanding in mathematics via problem solving. In P. R. Trafton & A. P. Shulte (Eds.), *New directions for elementary school mathematics. 1989 Yearbook* (pp. 31–42). Reston, VA: National Council of Teachers of Mathematics.

Thomas, M. O. J., & Chinnappan, M. (2008). Teaching and learning with technology: Realising the potential. In H. Forgasz, A. Barkatsas, A. Bishop, B. Clarke, S. Keast, W. -T. Seah, P. Sullivan, & S. Willis (Eds.), *Research in mathematics education in Australasia 2004–2007* (pp. 167–194). Sydney, AU: Sense Publishers.

Thomas, M. O. J., & Holton, D. (2003). Technology as a tool for teaching undergraduate mathematics. In A. J. Bishop, M. A. Clements, C. Keitel, J. Kilpatrick, & F. K. S. Leung (Eds.), *Second international handbook of mathematics education* (Vol. 1, pp. 347–390). Dordrecht, NL: Kluwer.

Thomas, M. O. J., & Hong, Y. Y. (2005a). Teacher factors in integration of graphic calculators into mathematics learning. In H. L. Chick & J. L. Vincent (Eds.), *Proceedings of the 29th conference of the international group for the psychology of mathematics education* (Vol. 4, pp. 257–264). Melbourne, AU.

Thomas, M. O. J., & Hong, Y. Y. (2005b). Learning mathematics with CAS calculators: Integration and partnership issues. *The Journal of Educational Research in Mathematics, 15*(2), 215–232.

Vérillon, P., & Rabardel, P. (1995). Cognition and artifacts: A contribution to the study of thought in relation to instrumented activity. *European Journal of Psychology of Education, 10*(1), 77–101.

Derek Holton
University of Otago (New Zealand)

Michael Thomas
The University of Auckland (New Zealand)

Anthony Harrdaine
Prince Alfred College (Australia)

TIM ROWLAND

4. JUST ENJOYING THE MATHEMATICS

INTRODUCTION

For many years I had the good fortune to be one of a team of lecturers teaching mathematics in Cambridge to undergraduate students who intended to become primary school teachers. The rationale for this curriculum for prospective teachers goes back to the 1963 Robbins Report, and is deeply embedded in my own thinking about teacher education (Rowland & Hatch, 2006). The courses we taught were 'proper' undergraduate mathematics. The content, and the pedagogical style to which many of us aspired, is indicated by the three 'Pathway' textbooks written by Bob Burn and published by Cambridge University Press (e.g. Burn, 1982). Alongside these courses, each with a content syllabus, sat a course with a process syllabus, albeit rather vaguely defined to begin with. In the early 1980s it was called, predictably, 'Investigations'. I was very keen to get my hands on this course, but had to be patient for 10 years until various people had left or retired, and I had the opportunity to teach it for the first time in 1992. By then it had been re-named Mathematical Processes (later still it became Mathematical Reasoning), and curtailed from two years (about 100 hours) to about 30 hours in the first term. In its final offering there were 12 hours. I should add that the assessment of each of the content-courses, such as Algebra and Geometry, included two (latterly, one) investigation-style projects, expected to take around 25 hours, within the domain of the taught course. Anyway, I was about to have 30 hours with these bright and mostly enthusiastic students, and I could do whatever I liked within the spirit of mathematical investigation.

I already had a kind of repertoire of 'starters' of my own, most of them gleaned from a variety of sources, but I was ever on the lookout for new ideas and suggestions. Essentially, my Mathematical Reasoning course consisted of a sequence of problems: I introduced them, the students solved them. There were shared reflections and discussions about how we had worked on the problems, and what we had learned by doing so – and what implications these might have for us as mathematicians and as teachers of mathematics. My own awareness of, and especially my ability to articulate, these Big Ideas for teaching and learning improved immeasurably as a consequence of trying to 'teach' this course. Particular instances of my own learning and awareness concern the distinction between inductive and deductive reasoning, and the explanatory power of generic examples (Rowland, 1999).

From the outset, however, I knew how I wanted the classes to proceed. Following some links with a previous session, I would introduce a problem situation, a

S. Lerman and B. Davis (eds.), Mathematical Action & Structures of Noticing: Studies on John Mason's Contribution to Mathematics Education, 49–61.
© 2009 Sense Publishers. All rights reserved.

'starter'. These tended to be 'pure', reflecting my own preference, with no problems about designing car parks and the like, although a problem might be set in a 'context'. Having introduced this starter, the students would work on them in pairs, threes at most, for around half an hour. It might be more or less, depending on their progress, and on their interest. In this phase, I made a point of not interacting with the pairs, unless I was asked to do so. This probably reflects my own dislike of "being watched" when I am working, alone or in a group. Although it is commonplace to 'circulate' and show interest in this phase, I really did not want to *interfere* in the exploratory work of my students, so I just left them alone. On the other hand, I needed clues about when to move on to a plenary discussion phase. More often than not, these came from overhearing the discussion of the pairs and triads. If I overheard a student say something especially interesting – in my eyes, of course – then I would forewarn the student that I would like to invite him/her to talk to the class about it later, and ask their permission to do so.

In the plenary, I usually selected two or three pairs to report their progress, then opened it to all comers. I joined in at this stage, and would occasionally pose a question to the class. Often this would invite a conjecture to be articulated. It was very clear, in most classes, that there were some students very eager to respond, and others very reluctant. One influential factor in this state of affairs, I believe, was that the Mathematical Reasoning course took place in their first term at university, at a critical time for students negotiating their role in the mathematics group. The 'pushy' ones more often than not turned out to be those keen to please and impress. They usually calmed down when they came to know that they were valued and accepted by their peers, and by their tutors. But there were also those who were reluctant to stick their heads above the parapet, afraid to be 'wrong' and to be thought foolish. Anne Watson once wrote that pupils can be

> ... worried about being wrong and nervous about asking for help if "being wrong" and "needing help" have, in the past, been causes of low self-esteem leading to ridicule, labelling or punishment. (1994, p. 6)

I made it my goal to try to create, with the students, what John Mason has called a 'conjecturing atmosphere'. Mason understood that some students are inhibited by social learning contexts: they might well have conjectures to offer, yet these might be tentative, speculative even. He wrote, therefore:

> Let it be the group task to encourage those who are unsure to be the ones to speak first ... every utterance is treated as a modifiable conjecture! (1988, p. 9)

I made a poster with these words, and displayed it in the classroom. I talked to the students about why I believed it to be an important insight, and an important guide for our conduct in this course. In fact I recommended three books (Mason et al., 1982; Mason, 1988; Pólya, 1945) as companions to the course, and directed the students to articles in professional journals (e.g. Rowland, 1999) from time to time. I don't think they paid much attention to these, until the time came for them to get down to the course assignment, which was to [attempt to] solve a problem and write a commentary on the process of solution. Later, I will show how Mason (1988) anticipated and supported what I was trying to do with these students.

JUST ENJOYING THE MATHEMATICS

Having set the scene, most of the remainder of this chapter will be an account of the mathematics developed in the course from three of the problem starting points. I include some of my own contributions and one from a 'third party'. My intention is to convey a sense of pleasure in the mathematics itself, in the surprise results and some unanticipated connections. Occasionally I indicate something of the heuristics in rather a general way, but sometimes I have just indulged myself, taking pleasure in just writing the mathematics.

THE FLOWERBED

This problem-situation was suggested to me by my former colleague Bob Hall[1] some time around 1995.

> A rectangular garden has a flowerbed in the middle. Between the boundary of the garden and the flowerbed is a path, the same width all the way round. The area of the flowerbed is half that of the garden. What are the possible dimensions of the flowerbed and the garden, if all the lengths (including the width of the path) are integers?

This statement is my own formulation, in fact. Until I checked, for the purpose of this chapter, I had completely forgotten that the original was formulated in terms of green and red tiles on a patio! When Bob first offered it to me, I just thanked him for the idea, but privately judged that it would not be very interesting (like G. H. Hardy and the taxi number 1729 perhaps). On the other hand, I knew that before long, Bob would ask me how I had got on with it.

I have no recollection of how I first 'solved' the flowerbed problem. Almost certainly, I drew a diagram like the one below:

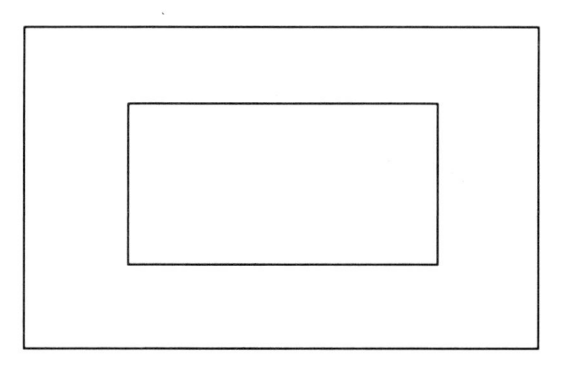

Figure 1. Solving the flowerbed problem.

... chose literal variables for the various dimensions, and expressed the relationships between them, as given in the problem, algebraically. In this way, I probably arrived at

$$(l + 2x)(w + 2x) = 2lw$$
$$2x(l + w) + 4x^2 = lw \tag{1}$$

Maybe I used a 'trick' to rewrite this as

$$(l - 2x)(w - 2x) = 8x^2 \tag{2}$$

Most likely I would then have found values of l, w and x to 'fit' this equation. Start with $x = 1$, so $(l - 2)(w - 2) = 8$, giving the following possible solutions:

$x = 1$	$l-2$	$w-2$	l	w	L	W
$(l-2)(w-2) = 8$	8	1	10	3	12	5
	4	2	6	4	8	6

Since equation (2) is symmetrical in l, w (i.e. invariant under transposition of these two variables), no essentially new solutions will result from interchanging the values of $l-2$ and $w-2$. [In other words, a solution remains a solution when I call the length the width, and vice versa].

Again, I don't remember if this was enough to prompt a conjecture. I would guess that it was enough: the L, W values are what 'do it' for me, especially the first pair (12, 5). If not, a couple of solutions with $x = 2$ should certainly be enough.

$x = 2$	$l-4$	$w-4$	l	w	L	W
$(l-4)(w-4) = 32$	32	1	36	5	40	9
	16	2	20	6	24	10

The second (L, W) solution here – (24, 10) – with border width 2, is double the earlier solution (12, 5) with border 1. If we double all the dimensions of any solution, the area of the flowerbed is still half that of the garden: and this would be true for any scaling of any solution. So, in that sense at least, solutions with x, l and w having no common factor are of particular interest. Such a solution will be referred to as a *primitive* solution.

In any case, it struck me pretty quickly that the (L, W) pairs (12, 5), (8, 6), (40, 9) and (24, 10) are the smallest elements of Pythagorean triples. This would not have occurred to me had I not been familiar with some Pythagorean triples. When I looked for the 'third element' in these same triples – 13, 10, 41, 26 respectively – I noticed something which struck me as even more remarkable. In each case, this third element is the sum of the l, w values. In other words, the two sides of the garden form a Pythagorean triple, with the half-perimeter of the flowerbed. Put differently, when the area of the flowerbed is indeed half the area of the garden, the half-perimeter of the flowerbed is equal to the diagonal of the garden.

This was an instance of one of those conjectures, based on very little empirical evidence, which I firmly believed to be true, because it was too remarkable to be mere coincidence. Because it was beautiful, it had to be true.

Beauty is truth, truth beauty – that is all
Ye know on earth, and all ye need to know.[2]

That belief did not lessen my determination to prove the conjecture!

JUST ENJOYING THE MATHEMATICS

The first proof was something like this:
I want to prove that $L^2 + W^2 = (l + w)^2$
Now $L^2 + W^2 = (l + 2x)^2 + (w + 2x)^2 = l^2 + w^2 + 4x(l + w) + 8x^2$

But equation (2), which expresses the 'given' concerning the area of the garden and the flowerbed, establishes the right hand expression to be $l^2 + w^2 + 2lw$, and so the proof is complete.

Returning to (2), it would be easier to solve if the number of variables was reduced. If the garden were square, so that $l = w$, we would have $(l - 2x)^2 = 8x^2$. Hence $l - 2x = 2\sqrt{2}x$, and this has no solution in integers since $\sqrt{2}$ is irrational. So there is no square solution. There is, of course, a square inside any given integer-side square with half its area: it's just that the sides of the smaller square are never integers.

In fact, given any rectangle with sides, L, W, if we contract the rectangle so as to leave the 'border' (x) between outer and inner rectangles the same, the rectangle shrinks continuously, so there must be a point at which its area must be ½LW (formally – by the Intermediate Value Theorem: informally – it's obvious). One could carry out this shrinking in a dynamic geometry environment, and measure the area at each stage. Now, we hit the half-area point when

$(L - 2x)(W - 2x) = ½ LW$, or $8x^2 - 4x(L + W) = -LW$

This is a quadratic in x, whose solution we wish to be an integer.
Doubling it gives $16x^2 - 8x(L + W) = -2LW$
Completing the square:

$$[4x - (L + W)]^2 = (L + W)^2 - 2LW \qquad (3)$$

So for x to be an integer, it is *necessary* that the right hand side, $L^2 + W^2$, be a perfect square – a nice proof of the Pythagorean triple conjecture.[3]

It struck me recently that both of these solutions invoke the border-width x, although it does not feature explicitly in the relationship that we are trying to prove, i.e. that $L^2 + W^2 = (l + w)^2$. Would it be possible to construct a proof that somehow bypassed the 'x'? My first effort began with the recognition that $2x$ is equal to both $L - l$ and $W - w$. Therefore $(L - l)^2 = (W - w)^2$ – but this would give me an expression for $L^2 - W^2$, whereas I wanted $L^2 + W^2$. It then occurred to me that $L - W = l - w$, although neither of these expressions corresponds to any identifiable dimension of the garden. Now

$(L - W)^2 = (l - w)^2$

Also $LW = 2lw$

Therefore $(L - W)^2 + 2LW = (l - w)^2 + 4lw$

Eureka!

Finally – as far as these proofs are concerned – it struck me soon after Bob Hall had introduced me to the problem, that all the proofs we had devised were algebraic. Yet the formulation of the problem is overtly spatial, about lengths and areas. Could we concoct a proof that was somehow geometrical at heart? The answer, for myself and Bob at least, was 'no', we could not! In the end, Bob referred the problem to his brother, Alf. Alf was a librarian, with no formal mathematical training beyond school, who liked 'dabbling in these things'. Within a few days, the solution came in a letter from Alf, in his neat copperplate hand. It began "Dear Rob, On looking at the rectangle problem again, I found a concise geometric proof of the theorem ... the big red herring, from the geometric point of view, is to align the two rectangles with a constant border all round. Move the inner rectangle into the corner of the outer one, and all becomes clear." Alf's argument is quite dense from that point. It is purely narrative – not one literal variable or equality sign is used, an object lesson to students who believe that 'proper' proofs must look algebraic. Here is my version of his proof:

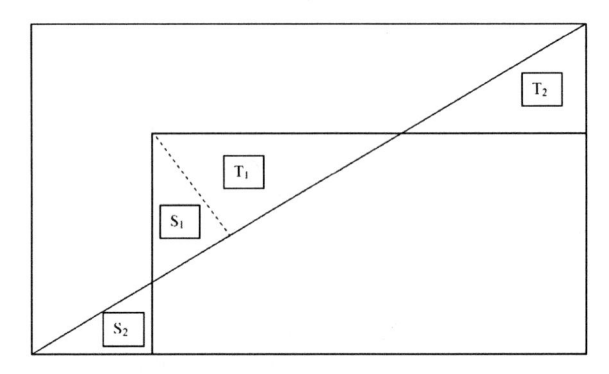

Figure 2. My version of Alf's proof.

Construct the (dotted) altitude of the 'central' triangle between the smaller rectangle and the diagonal of the larger one. Note that (a) the diagonal partitions the larger rectangle into two triangles of equal area, *and* (b) the area of the smaller rectangle is equal to that of the remaining strip of the larger rectangle. It follows that the area of the central triangle is equal to the sum of the areas of the two corner ones. Now T_1 is similar to T_2, and likewise S_1 and S_2, *and* in the same ratio (dotted altitude: remaining strip width). It follows that this ratio is equal to 1, and the similar triangles are actually congruent. But the corresponding edges of T_1 and T_2 interchange segments of the sides of the smaller rectangle with segments of the diagonal of the larger rectangle. Likewise with S_1 and S_2. Tracking these segments shows that the sum of the sides of the smaller rectangle equals the diagonal of the larger rectangle.

The answer to the question: what is the general solution of the flowerbed problem? – seems somewhat prosaic after that. Anyway, here it is:

We know that every solution corresponds to a Pythagorean triple, but the converse is not true. Given a primitive Pythagorean triple, one can try to *construct* a solution $(L, W, l + w)$ of the flowerbed problem. For example, the triple (33, 56, 65) suggest $L = 33$, $W = 56$, and then $(33 + 56) - 65 = (L - l) + (W - w) = 4x$, so $x = 6$ and $l = 21$, $w = 44$ (with $l + w = 56$ as expected). It's easily checked that $LW =$

1848 and $lw = 924$. On the other hand, applying the same steps to the triple (8, 15, 17) gives the non-integer border $x = 1\frac{1}{2}$. However, doubling the dimensions, (16, 30, 34) gives $x = 3$ and $l = 10$, $w = 24$.

In general, every Pythagorean triple is form $(k(2mn), k(m^2 - n^2), k(m^2 + n^2))$, where k, m, n are positive integers; m, n are coprime and of opposite parity; and $m > n$. As above, this will yield a solution $(L, W, l + w)$ of the problem, with $4x = L + W - (l + w)$. For x to be an integer, $kn(m + n)$ must be even. Now $m + n$ is odd (opposite parity) so therefore either k or m must be even.

If n is even, then k must be 1 for $(k(2mn), k(m^2 - n^2), k(m^2 + n^2))$ to yield a primitive solution.

Let n be odd. Then k must be even, and if the corresponding solution $(L, W, l + w)$ is to be primitive, $k = 2$.

In conclusion: every primitive (L, W, x) is of one of the following two forms:

$L = 2mn$, $W = m^2 - n^2$ (or vice versa), $x = \frac{1}{2}n(m + n)$,
where m, n are coprime, m is odd, n even, and $m > n$

$L = 4mn$, $W = 2(m^2 - n^2)$ (or vice versa), $x = n(m + n)$,
where m, n are coprime, m is even, n odd, and $m > n$

STAIRS AND PARTITIONS

> **Stairs**: In how many different ways can you ascend a flight of stairs in ones and twos?
>
> **Partitions**: The number 3 can be 'partitioned' into an ordered sum of (one or more) positive numbers in the following four ways: 3, 2 + 1, 1 + 2, 1 + 1 + 1. In how many ways can other positive numbers be partitioned?

These two combinatorial problems are connected, though I tended to offer Stairs to the students first. So I'll start with that one.

The sequence of numbers-of-ways is 1, 2, 3, 5, 8, 13, ... Students usually readily recognise what they call "the pattern", and many have encountered the Fibonacci sequence before. The interesting bit is explaining *why* each term is the sum of the previous pair. Invariably, we come at it through a generic example: sometimes a student proposes one spontaneously, or I offer a prompt by inviting them to consider how you could list some ways of ascending 6 stairs, based on the list of ways for 5 stairs. In the end a generic argument looks something like this:

Suppose we've done lengths 1 to 5 by listing all the ways. Now consider ascending 6 stairs. The first step must be 1 or 2. If it is 1, then 5 stairs are left, and we know that there are 8 ways to ascend them. If it is 2, then 4 stairs are left, and we know that there are 5 ways to ascend them. This exhausts all the possibilities, so there are 5 + 8 ways of climbing the 6 stairs.

Over the years, I've strenuously advocated the pedagogic use of proof by generic example, and made tentative proposals for the construction of such

55

examples. I was greatly impressed and influenced by Mason and Pimm (1984), which, when I first read it some ten years after publication, was a Damascus Road experience. It gave a name to something I had come to recognise without knowing it had a name, and a literature. I won't rehearse the evidence again here, but year-on-year, students have found the generic argument a helpful, and a natural, way into the general. The transition between generic ways of knowing and the writing of general proofs is more a matter of notation than of comprehension.

Returning to the stairs problem: when I ask what would happen if you could take in 1, 2 or 3 stairs each time, students very readily make the expansive generalisation of the Fibonacci sequence. This is made possible because they have 'seen' why the previous pair were summed in the original version of the problem.

Among the multiple approaches to the stairs problem proposed by students over the years, I found the following one particularly interesting. Consider 7 stairs (generic again ...). If every step is a '1', then there is just 1 way to do it. If one step is a '2', then five steps are 1s so there are 6 elements, and the 2 can be any one of them (e.g. 211111, 121111 etc). So there are 6 ways with just one 2. Now if two steps are 2s, there are five elements altogether (three 1s, two 2s) and there are $^5C_2$ ways of assigning the two 2s in the sequence of five steps. Similarly if three steps are 2s, there are four elements and $^4C_3$ ways of assigning the 2s. Thus the total number of ways for 7 steps is $1 + {}^6C_1 + {}^5C_2 + {}^4C_3$. This equals $1 + 6 + 10 + 4 = 21$, a Fibonacci number, as we know it should be. Likewise, the total number of ways for 8 steps will be $1 + {}^7C_1 + {}^6C_2 + {}^5C_3 + 1$ (there being just one way if all four steps are 2s).

Now these binomial coefficients are the terms in the diagonal lines on a Pascal's triangle (fig. 3, shown in black, dark grey respectively, with light grey for 5 steps).

It's rather pleasing to see that the sum of each of these diagonals is a Fibonacci number – something I didn't know before.[4] A more direct proof of this property, by mathematical induction, follows from the way that each row of Pascal's triangle is usually constructed from the previous row.

And so to the Partitions problem: note that, unlike 'partitions' in number theory or combinatorics, these are *ordered* partitions. A doubling pattern 1, 2, 4, 8, is readily recognised, and a prediction of 16 ways for the number 5 readily made and – usually – carefully checked. Conjectures about doubling, or powers of 2, turn out to be easy to state but harder to prove. More often than not, I offer a generic hint such as: list all four partitions of 3, in a column say, and all the eight partitions of 4 in a second column. If you knew all the partitions of 3, how could you immediately list four partitions of 4? (Note: this confusion of the role of '4' is not ideal, but I don't want them to have to list all the partitions of 5!). What about the other four partitions of 4? How do they relate to your four partitions of 3?

Typically, before too long, someone articulates an argument such as: suppose we've listed all of the 8 partitions of 4. Consider any one of them, e.g. $2 + 1 + 1$. If I adjoin an additional part of size 1 (to make $2 + 1 + 1 + 1$), I have produced a partition of 5. If, instead, I augment the size of the last part by 1 (to make $2 + 1 + 2$), I have produced a second partition of 5. In this way, every partition of 4 yields two partitions of 5, so the number doubles.

JUST ENJOYING THE MATHEMATICS

1								
1	1							
1	2	1						
1	3	3	1					
1	4	6	4	1				
1	5	10	10	5	1			
1	6	15	20	15	6	1		
1	7	21	35	35	21	7	1	
1	8	28	56	70	56	28	8	1

Figure 3. Pascal's Triangle and the Fibonacci Numbers.

Actually, this argument established that the number of partitions of 4 is *at least* double that for 3. The final step in the argument is to show that there is no partition of 4 that does not result from a partition of 3 and the application of one of the two processes described above. Well, every partition of 8 'ends' in 1, or not. In the first case, if the final 1 is removed, we have partition of 3. In the second case, reducing the final part by 1 takes us back to a partition of 3.

The smart proof, of course, is to envisage an integer n as a rod of length n, like a Cuisenaire rod, with $n - 1$ notches marking the unit intervals of the rod (note the power of certain *representations*). A partition then consists of a set of saw-cuts through the rod at some (possibly all, or none) of the $n - 1$ notches. At each notch, there is a choice of making a cut, or not, and each two-option choice is independent of the other choices. Hence there are 2^{n-1} possible partitions.

Finally, a nice connection between Stairs and Partitions is the following. The stairs problem is equivalent to partitioning each positive integer into parts of size not greater than 2. As we hinted earlier, if you could take in 1, 2 or 3 stairs each time, the number of ways of ascending n stairs is the sum of the numbers of ways for the previous three stair lengths: and this is equivalent to partitioning each positive integer into parts of size not greater than 3. If, then, we allow partitions into parts of *any* size, equivalent to Stairs and steps of any length, then the number of ways of partitioning n will be the sum of the ways of partitioning *all* smaller positive integers. If the number of partitions of n is p_n, then $p_{n+1} = p_n + p_{n-1} + \ldots + p_1$. Yet this clearly isn't quite right. If it were, then p_3 would be 3 (= 2 + 1) and not 4. This is because the argument for the Fibonacci-type recursive relation for Stairs supposes that the permissible step lengths for $n + 1$ stairs will be the same as those for shorter staircases (e.g. 1 or 2 in the original problem). But in the partitions problem, a partition with just one part of size $n + 1$ is permitted, whereas $n + 1$ did not feature in any of the earlier partitions. The recurrence relation must therefore be modified to: $p_{n+1} = p_n + p_{n-1} + \ldots + p_1 + 1$. This agrees, in a satisfying way, with the well-known geometric series sum: $1 + 2 + 2^2 + \ldots + 2^{n-1} = 2^n - 1$.

CONCLUSION

This chapter has been a kind of tour around some possible approaches to solving three problems, demonstrating both the diversity of possible approaches and the links, sometimes unexpected, between the problems and the solutions. I have enjoyed doing the mathematics again, organising a file of disorganised field notes, tracking the genesis of some of the solutions, and setting them down on paper. But the reason for working on the problems in the first place was to draw a class of prospective teachers into mathematical activity of a non-passive kind, where there were no notes to copy and no theorems and proofs to be learned for an examination. There were things to be learned, however, to do with awareness of what human beings do when they solve problems, when they bring new mathematics into being. Without those awarenesses, the mathematics student (and the mathematics teacher, for that matter) operates as a kind of automaton, making trained responses to situations without ever being in control of them, or of him/herself. Caleb Gattegno wrote that "only awareness is educable", an insight which came to me via Mason (1987). I itemise here a few of those awarenesses, most of which are discussed at length in those two recommended books (Mason, 1982, 1988).

The nature of inductive reasoning and the crucial importance of conjectures

New mathematics comes into being when (though not only when) regularities are observed and articulated as conjectures. A conjecture is a 'maybe' statement about a whole class of things on the evidence of just a few of them, sometimes only one. A crucial step in the solution of each of the problems discussed earlier was the tentative articulation of a conjecture, such as "it doubles each time". Given that the one who states a conjecture can never know that what they say is true, John's advice to "encourage those who are unsure to be the ones to speak first" is especially apposite. Given this very particular meaning of the word 'conjecture', I urge students to use it to describe those particular insightful, but provisional, uncertain products of their mathematical activity, in preference to words like 'theory' and 'hypothesis' that seems to come to them more readily. Awareness of 'having' a conjecture is a cause for excitement and anticipation. The public articulation of a conjecture is a gift to the group or the class. It becomes a 'thing' out there to be entertained in a conjecturing atmosphere, to be tested and possibly refuted, perhaps to be proved.

JUST ENJOYING THE MATHEMATICS

The various forms and purposes of proof

Several writers have itemised the purposes of proof within mathematics. These include verification, conviction and systematisation. Undergraduate texts convey the importance of the first and third of these, at the expense of the second. But then such texts only ever set out to prove statements that are already known to be true (why else publish them in the textbook?) and so unwittingly uphold the belief that the student's task is to assimilate someone else's (usually deceased males) mathematics, not to write their own. Mason, Burton and Stacey (1982, p. 95) famously address the second purpose, and the significance of the *audience* of a proof-demonstration.

Convince yourself; convince a friend; convince an enemy

This function of proof – to convince one who lacks conviction – is paramount when the statements to be proved are not great Theorems from the canon of mathematics, but local conjectures, provisional truths, generated by students. No amount of additional confirmation will suffice: the occasion now demands that an explanation, a deductive argument be assembled. First, convince yourself, says Mason. If you can do that, convince a friend next, then convince an enemy. I am bound to say that we rarely got beyond the second stage, because the spirit was one of affirmation rather than demolition. It was necessary to remind students that there are no restrictions on what might 'count' as a proof, especially when it has to succeed in convincing another person. The students' tendency was to expect that narratives, thought-experiments and diagrams were poor substitutes for strings of symbols, however obscure (and often meaningless: witness the liberal use of the implication symbol as a kind of connecting tissue between various algebraic expressions). Evidently nothing in the previous three-to-four years had disturbed the beliefs that they most likely held as high-achieving 15-year-olds (Healy & Hoyles, 2000). I often found that students' initial, faltering attempts to construct a convincing, general argument would begin from a critical analysis of a particular case – a generic example. Given my own Damascus Road conversion to this mode (e.g. Rowland, 1999) I made much of it as a proof-mode in its own right, and as a stepping-stone to arguments that make no explicit reference to particular cases.

The place of examples in mathematics

The importance of examples and exemplification in mathematical activity cannot be exaggerated. Through particular examples, we both construct and make sense of general concepts, procedures and truths. Most of the time, maybe all of the time, mathematics is about generalising and specialising, themes that run throughout Mason's writing and teaching. I was pleased to discover the final interlude: *On Examples*, in Mason (1988) recently. I seem to have been saying some of the same things myself recently, and suspect that a citation would have been in order. The notions of dimensions of variation, dimensions of possible variation, and permissible change within each dimension, are valuable ways of identifying and testing the scope of both problem-situations and consequent conjectures.

Having said all this – it would be a kind of delusion to claim, or even to believe, that all the students who attended my 'processes' course went away with a born-again, revised view of mathematics and mathematical activity, although, at a superficial level, some of them attested to the usefulness of a repertoire of potentially-productive problem starters when they began to teach. There is evidence that what I regarded as some of the key ideas of the course were poorly understood by some students. In submitted coursework, for example, mere confirming instances of conjectures were sometimes offered as generic examples, without the structure that might make them a form of proof. Of course, this was how I interpreted what they had written. There was evidence elsewhere that the 'penny' had not always dropped as intended. In the end-of-first-year mathematics examination there was always one question on 'mathematical reasoning'. This was anomalous in that we always argued that the course was best assessed by coursework. On the other hand, including such a question in the 'proper' mathematics papers gave status to the 'processes' course, or so we hoped. The 2000 examination included the following question.

> Explain the difference between *inductive* and *deductive* reasoning, and illustrate by detailed reference to <u>one</u>[5] of the following:
>
> an enquiry into which numbers can be expressed as a sum of two or more consecutive natural numbers (your account should be selective and not exhaustive);
>
> Wilson's theorem that $(p-1)! \equiv (p-1) \bmod p$, for all prime numbers p.

Several answers indicated that my supposedly carefully-made distinction between the two kinds of reasoning, exemplified and celebrated at numerous points in the course, was not universally recognised, accepted or acclaimed. A few students even wrote about Proof by Induction, when I had been at pains to explain that the 'new' intended meaning of 'induction' was at odds with the only usage – as a rather specialised mode of proof – which most of them would have encountered before coming to university. I contrast the intended ideal with the reality here as a way of reminding myself, if it were necessary, that students are capable of engaging (indeed, engaging productively) in acts of specialising, generalising, conjecturing, convincing and being convinced, without stepping outside the acts themselves sufficiently to become aware of them: or, at least, to become aware of the ways that language is being used to refer to what they have been doing. Other students seem to develop these same awarenesses readily, and to reflect and act upon them. In the end, I believe that they all constructed the course, and made sense of my dozen-or-so hours with them, in their own distinct and individual ways. And so did I, as I wrote this account of what we did together.

NOTES

[1] For a taste of Bob's inventiveness, see Hall (1992), for which I acted as a kind of amanuensis ...
[2] John Keats, *Ode to a Grecian Urn*.
[3] This proof is a variant of that proposed by a student, Catherine Benoist, in 1999.
[4] Although others evidently did – e.g. http://goldennumber.net/pascal.htm
[5] Neither of these situations was being offered to the candidates 'cold'. The first had been the subject of class investigation in the Processes course that year; the second had been 'discovered' and proved in a course on Number Theory.

REFERENCES

ATM. (1967). *Notes on mathematics in primary schools*. Cambridge, UK: Cambridge University Press.
Burn, R. (1982). *Pathway into number theory*. Cambridge, UK: Cambridge University Press.
Hall, B. (1992). Gauss's counter example. *The Mathematical Gazette, 76*(477), 359–361.
Healy, L., & Hoyles, C. (2000). A study of proof conceptions in algebra. *Journal for Research in Mathematics Education, 31*(4), 396–428.
Mason, J. (1987). Only awareness is educable. *Mathematics Teaching, 120*, 30–31.
Mason, J. (1988). *Learning and doing mathematics*. London: Macmillan.
Mason, J., Burton, L., & Stacey, K. (1982). *Thinking mathematically*. Wokingham, UK: Addison Wesley.
Mason, J., & Pimm, D. (1984). Generic examples: Seeing the general in the particular. *Educational Studies in Mathematics, 15*(3), 277–289.
Pólya, G. (1945). *How to solve it: A new aspect of mathematical method*. Princeton, NJ: Princeton University Press.
Rowland, T. (1999). 'i' is for induction. *Mathematics Teaching, 167*, 23–27.
Rowland, T., & Hatch, G. (2007). Learning to teach? The assistant lecturer in colleges of education 1960–75. *History of Education, 36*(1), 65–88.
Watson, A. (1994). My classroom. In A. Bloomfield & T. Harries (Eds.), *Teaching, learning and mathematics* (pp. 6–8). Derby, UK: Association of Teachers of Mathematics.
Watson, A., & Mason, J. (2005). *Mathematics as a constructive activity: Learners generating examples*. Mahwah, NJ: Erlbaum.

Tim Rowland
University of Cambridge (United Kingdom)

OLIVE CHAPMAN

5. SELF-STUDY AS A BASIS OF PROSPECTIVE MATHEMATICS TEACHERS' LEARNING OF PROBLEM SOLVING FOR TEACHING

INTRODUCTION

Problem solving is promoted as a way of doing, learning and teaching mathematics (Kilpatrick, Swafford, & Findell, 2001; NCTM, 1989, 2000). However, whether or how such ways of viewing problem solving get implemented in the classroom will depend on the teacher. If problem solving should be taught to students, then it should be taught to prospective teachers who are likely to enter teacher preparation programs without having been taught it in an explicit way. If it is to form a basis of teaching mathematics, then prospective teachers should also understand it from a pedagogical perspective. In this chapter I discuss an instructional approach to facilitate prospective teachers' learning of problem solving for teaching, in particular, their development of understanding of problems, problem solver, problem-solving process, problem-solving pedagogy and problem solving as inquiry-based teaching. I reported the initial version of this approach in Chapman (2005). This version extends and further enhances the approach. The chapter provides a description of the approach, its relationship to John Mason's work in terms of the theoretical perspective framing it, the results of an investigation of its effectiveness, and implications for teacher education.

RELATED LITERATURE

Some studies on prospective teachers have raised issues about their knowledge of problem solving for teaching. For example, Verschaffel, De Corte, and Borghart (1996) investigated 332 prospective teachers' conceptions and beliefs about the role of real-world knowledge in arithmetic word problem solving. For each of the 14 word problems, the prospective teachers were first asked to solve the problem themselves, and then to evaluate four different answers given by students. The results revealed a strong overall tendency among the participants to exclude real-world knowledge and realistic considerations from their own spontaneous solutions of school word problems as well as from their appreciations of the students' answers. Taplin (1996) explored the approaches to problem solving used by 40 prospective elementary teachers and found that they preferred to work with a narrow range of strategies, predominantly verbal and numerical. They tended to select a method of approach and not change from that through the tutorial, implying inflexibility in their choice or management of problem-solving strategies. Chapman (2005) examined 26 prospective secondary mathematics teachers'

S. Lerman and B. Davis (eds.), Mathematical Action & Structures of Noticing: Studies on John Mason's Contribution to Mathematics Education, 63–73.
© 2009 Sense Publishers. All rights reserved.

knowledge of problems and problem solving and found that, in relation to their initial knowledge, most of the participants made sense of problems in terms of the traditional, routine problems they had experienced prior to entering the teacher education program. They understood the problem-solving process in a way consistent with the traditional classroom approach of dealing with routine problems. Studies such as these imply concerns about how prospective teachers may conceptualize problem solving and engage in it. They also suggest that prospective teachers are likely to need help in their development and understanding of problem solving from the perspectives of a learner and a teacher.

Some studies have investigated instructional practices that directly or indirectly addressed helping prospective teachers to grow in their knowledge of problem solving for teaching. For example, Lee (2005) studied an approach to help prospective teachers develop in their role of facilitating students' mathematical problem solving with a technology tool. The approach consisted of a cycle of "planning–experience–reflection," which was repeated twice during a course to allow the prospective teachers to change their strategies when working with two different groups of students. The phases of the cycle consisted of the prospective teachers individually solving an open-ended problem and planning a learning trajectory for students; interacting with two students as they solved the same problem; discussing the experience with peers, planning a revised learning trajectory for different students and reflecting on their role in facilitating students' problem solving and their understanding of what the students understood about the problem. The effect of the approach, based on a study of three students, was that the planning–experience–reflection cycle provided opportunities for them to begin to struggle with issues of facilitating students' problem solving and to make their struggle an open and reflective activity used as an opportunity to improve their practice. In Szydlik, Szydlik, and Benson (2003), the implied goal of this approach in relation to problem solving was to help the prospective teachers to become autonomous problem solvers by promoting community autonomy rather than individual autonomy. In this approach, the students worked on "demanding problems" in small groups, then discussed their findings, strategies, solutions and arguments. The instructor provided no feedback on the correctness or completeness of any solution, almost no assistance in the problem solving aspect of the course, and no answers for the problems. The only way for the participants to understand a problem was to figure it out. The only way to know they were correct was to find a convincing argument. The authors concluded that a classroom focusing on problem solving using a variety of strategies, reflection on the process, and engagement in the process of exploration, conjecture, and argument can help prospective teachers develop mathematical beliefs that are consistent with autonomous behavior. The participants experienced a broadening in the acceptable methods of solving problems and became more supportive of autonomous behaviors during the course.

A central aspect of these approaches is engaging prospective teachers in problem solving and reflection. In Lee's approach, reflection was a means for them to learn about the problem solver and how to facilitate his/her learning of problem solving using technology. In Szydlik, Szydlik, and Benson's case, it was a means for them to make sense of their problem-solving processes. Implied is also a self-study component in which the prospective teachers learned from their own experiences through reflection and collaboration with peers. While these studies were not intended to emphasize the notion of self-study as a basis of prospective teachers' learning of mathematical problem solving, in this chapter it is treated as such. Thus the study being reported on offers an approach (the *Self-Study Approach*) that highlights self-study and personal experience as a basis of learning.

PERSONAL EXPERIENCE AS A BASIS OF LEARNING

"Self-study" is used here to refer to the study of one's thinking and actions, to gain new understandings of oneself and situations relevant to one's experience. The Self-Study Approach allows prospective teachers to study themselves, directly or indirectly, to acquire knowledge of non-routine problem solving. It emphasizes personal experience as a basis of learning. This is related to a theme in Mason's work (Mason, 1994, 2002; Mason, Burton, & Stacey, 1982), which will be used to describe the theoretical underpinnings of the Self-Study Approach. This approach also deals with mathematical thinking as an integral aspect of non-routine problem solving. A summary of related notions of thinking mathematically and learning from personal experience based on these Mason's references are provided next, followed by a brief description of how they relate to the Self-Study Approach.

Thinking mathematically

Mason, Burton, and Stacey (1982) addressed mathematical thinking in relation to non-routine problem solving and more generally as "a dynamic process which ... expands our understanding" (p. 158). They explained that it is supported by "an atmosphere of questioning, challenging, reflecting" (p. 159) and provoked by "a challenge, a surprise, a contradiction, a perceived gap in understanding" (p. 159). It leads "to a deeper understanding of yourself. To a more coherent view of what you know. To a more effective investigation of what you want to know" (p. 159). It "helps in understanding yourself and the world" (p. xi). It "can be improved by tackling questions conscientiously; reflecting on this experience; linking feelings with action; studying the process of resolving problems and noticing how what you learn fits in with your own experience" (p. ix). This suggests that mathematical thinking can be improved by learning from personal experience.

Learning from personal experience

Mason's work emphasizes, directly and indirectly, the importance of learning from one's own experience; "to talk *from* experience rather than talking *about* it" (Mason, 1994, p. 177). As he explained, "experiential learning is based on learning from experience, but this requires more than mere experiencing" (Mason, 2002, p. 29). It also requires practices involving "researching from the inside" (Mason, 1994) and "noticing" (Mason, 2002).

Researching from the inside, as a basis of learning for teachers, according to Mason, requires that teachers examine their own experience of work on themselves while addressing the question of how to support students in learning mathematics. They have to work on themselves, informed by research and shared practice. Mason explained, "The core of researching from the inside is attending to experience ... so as to develop sensitivities to others and to be awake to possibilities" (Mason, 1994, p. 180).

Noticing, as a basis of teachers' learning, "is a collection of practices both for living in, and hence learning from, experience, and for informing future practice" (Mason, 2002, p.29). It is "a reference to lived experience through an invitation to check something out in your own experience" (Mason, 2002, p. xi). It is also an important component of *researching from the inside*. Two key factors that define the uniqueness of these bases of teachers' learning are

> *introspective observation* (in which an inner witness observes the self caught up in the action, yielding inner objectivity experienced subjectively); and *interspective observation* (in which people share observations as witness to each other, yielding objectivity from negotiated subjective information). (Mason, 2002, p. 85)

In addition, *noticing,* as a collection of systematic practices to support and enhance learning, consists of four interconnected actions (Mason, 2002, 1994):

> *Preparing and Noticing:* entering recent experiences post-spectively, then imagining making an alternative choice prospectively (1994, p. 183); reflecting on the past by reentering situations as vividly as possible and preparing to notice in the future by imagining oneself choosing to act. (2002, p. 93)

> *Systematic Reflection:* retrospective re-entry through brief-but-vivid accounts of incidents, without judgement or explanation (1994, p. 183); working in them [accounts] so that others recognize something from their own experience; developing sensitivities by seeking threads among those accounts, and preparing oneself to notice more detail in the future. (2002, p. 93)

> *Recognising and Labelling Choices*: locating alternative strategies and gambits for use in particular situations in the future, either from reading, from sharing incidents, or from observing others (1994, p. 184); being on the lookout to notice alternative behaviours or acts ... ; labeling salient incidents and alternative acts so that they begin to form a rich web of interconnected experiences. (2002, p. 93)

> *Validating with Others*: continued interspective exchange of brief-but-vivid incidents and observations (1994, p. 184); merging of the world of personal experience; the world of one's colleagues' experience; the world of observations, accounts, and theories, which informs what might be noticed, structures what is noticed. (2002, p. 93)

This suggests, then, that an effective self-study approach should include relevant aspects of these actions to facilitate deep learning based on one's past, current and future experiences.

The Self-Study Approach

The Self-Study Approach was developed to help prospective secondary mathematics teachers in an inquiry-based, student-focused, teacher preparation programme to develop knowledge of problem solving for teaching. The underlying principle of this approach is for the participants to examine their own experiences of work on themselves and comparing it with others and theory. Thus, the approach consists of activities that require them to engage in aspects of both *introspective* and *interspective* observations, for example, attending to their own experiences, and that of others, with solving non-algorithmic problems. In general, the activities require participants to engage in aspects of *mathematical thinking, researching from the inside* and *noticing*. Some of these aspects are highlighted next.

The Self-Study Approach consists of three stages of activities. The first stage focuses on self-reflection on mathematical problems and problem solving in order to create awareness of prospective teachers' initial conceptions and knowledge. It requires participants: (i) to reenter the past directly by describing, or indirectly by applying what they know about, problems and problem solving based on past experiences; (ii) to imagine and make alternative problems to those of their own experiences as students of mathematics; and (iii) to engage in systematic reflection of their experiences, recognize and label characteristics of problems and problem solving and validate these characteristics by working with peers to compare experiences. The second stage consists of inquiry activities intended to modify and extend the prospective teachers' initial conceptions and knowledge. Participants are expected to work on all problems in these activities without the facilitator's intervention in order to allow them to learn from their experiences with the activities. This stage requires them: (i) to enter alternative experiences to those of their past experiences with problems and problem solving; (ii) to engage in systematic reflection of their experiences; (iii) to recognize and label characteristics of problems and problem solving; and (iv) to validate outcomes of their inquiries by working with peers to compare experiences. The third stage includes activities that require the prospective teachers to compare their post-Stage 2 thinking with their pre-Stage 2 thinking and to validate their thinking by comparison to theory. The next section summarizes the activities for each stage.

THE SELF-STUDY APPROACH ACTIVITIES

The following is a description of the key activities of this Self-Study Approach.

Stage-1 Activities:

Participants respond to a list of questions that includes: What is a problem? Choose a grade at the secondary school level and create a mathematics problem that would be a problem for those students. What did you think of to create the problem? Why is it a problem? Is it a 'good' mathematics problem? Why? What process do you go through when you solve a problem? If possible, represent the process with a flowchart.

CHAPMAN

Stage-2 Activities:

These consist of four tasks.

(1) Comparing different types of "problems" (Table 1) without solving them in order to explore the nature of problems used in teaching mathematics and the goal of these problems in learning mathematics. The first four "problems" are from textbooks used in local schools.

Table 1. Problems to compare

---

1) Simplify: $\dfrac{-3(a-1)-2a}{-3+5a}$
2) The receipts from 550 people attending a play were $9184. The tickets cost $20 for adults and $12 for students. Find the numbers of adult tickets and student tickets sold.
3) Dylan is meeting his sister and four of her women friends for lunch. The five women are called Alicia, Rachel, Lani, Donna, and Casey. Three of the women are under 30 years old, and two are over 30. Two of the women are lawyers, and three are doctors. Alicia and Lani are in the same age group. Donna and Casey are in different age groups. Rachel and Casey have the same profession. Lani and Donna have different professions. Dylan's sister is a lawyer and is over 30. Who is Dylan's sister?
4) A road up one side of a hill is 12 km long, and it is 12 km down the other side. Suppose you can cycle up the hill at 6 km/h. How fast would you have to cycle down the other side to average 12 km/h for the entire trip?
5) How much paper of all kinds does your school use in a month?
6) Here is the number seven written as a Roman numeral with four toothpicks.

   \/||

   Change the position of only one toothpick to get the number one. (Note: all four toothpicks must be used.)

---

(2) Writing a narrative of the experience of solving a non-algorithmic problem, then analyzing the narrative in terms of the cognitive and affective aspects of the experience, e.g., examining what happened when "stuck" and considering when and how teacher intervention would be meaningful. The narrative has to be a temporal account not only of the mental and physical activities to solve the problem, but the emotional aspects of the experience. To illustrate this, Table 2 provides a portion of a participant's narrative for the following problem.

> *Emma was always looking for ways to save money. While in the remnant shop she came across just the material she wanted to make a tablecloth. Unfortunately the piece of material was in the form of a 2 m × 5 m rectangle and her table was 3m square. She bought it however having decided that the area was more than enough to cover the table. When she got home however she decided she had been a fool because she couldn't see how to cut up the material to make a square. But just as she despaired she had a brainwave, and with 3 straight cuts, in no time at all, she had 5 pieces that fitted neatly together in a symmetric pattern to form a square using all the material. How did she do it?*

Table 2. Problem solving narrative

... I anticipated that the math problem would take a long time. I ... initially assumed that it would be a simple problem involving perimeter or area. At this point in time I was feeling very confident and was not intimidated by the length of the problem. As I read through the problem I attempted to visualize it. However, the more I read the more confusing it became to visualize. In my mind it was a lot of information, so I decided to write down what I was given, and what I needed to find.

I knew that Emma's table had dimensions 3 m by 3 m and a total area of 9 m square. I also knew that the piece of material was a rectangle with dimensions 2 m by 5 m and a total area of 10 m square. The question was also explicit in saying that Emma made 3 straight cuts which created a total of 5 pieces that fit together in a symmetrical pattern to make a square. I took a moment to clear my mind and I looked at the information I had. I proceeded to draw the material and Emma's table.

I stepped back and attempted to look at the problem geometrically.

[...]

I thought that I had made the problem easier by minimizing the material that I had to work with, however, it was exactly the same problem in a different manner. I could feel myself tensing up because I had run into a dead end. How could I fit a 4 m square fabric into a 3 m square space? I was so involved in the problem that I was fighting the signs that my stomach was sending to me. I needed to eat so I decided to take a small break and come back with a clear mind.

I came back approximately 30 minutes and started that problem again.

[...]

I had been working on this question for about 1 hour and was starting to lose interest, but I kept going with the hope that I would eventually arrive at the solution. I had approached this question initially with such confidence; however, I now felt a little embarrassed by my inability to solve it in a short amount of time.

[...]

I wanted to give myself time to review the steps I had taken, that had brought me to my current situation. At this point in time I had 5 pieces altogether in which 4 were triangles with dimensions 1m by 2 m each, and 1 rectangle with dimension 2 m by 3 m. I decided to put the 2 m by 3 m rectangle directly on top of the square table such that the bottoms of each figure lined up. While using the cut outs I made from paper I noticed that it wasn't going to work out. ... I started to doubt myself and my ability to solve this problem. I became frustrated and could not decide what to do next. I stopped every thing, closed my eyes, and took a deep breadth. I was beginning to feel inadequate, and wondered to myself if I would become a good math teacher.

I decided to read the question all over again. This time I decided to focus on just the rectangular fabric and see if I could get 5 pieces from 3 straight cuts. I tried to cut the rectangular fabric vertically, horizontally, and diagonally while attempting to keep a sense of symmetry. I was able to obtain symmetry, however, I could not make 5 pieces and vice versa. I started to question the information I had and wondered if this was even possible. I began reading into the problem and more information came up in my mind:

1) How can a rectangular fabric of area 10 m square completely cover a 9m square table without having any overlap or overhang?
2) What does the problem mean by 3 straight cuts? Does it include cuts that are diagonal?
3) Do the five pieces form a square that is only one layer thick?
4) Is the square going to be smaller than 9 m square, equal to 9 m square, or larger than 9 m square?
5) Suddenly, a problem that had initially seemed simple became nearly impossible.

[...]

I wasn't sure if I had the right answer, but from where I saw it I already do. I felt as if a weight had been lifted from my shoulders. I was smiling and the stress was gone. The journey that I traveled on to reach this point brought much frustration and doubt. However, it became all of it, with the satisfaction of knowing that I had done it.

(3) Investigating others (peers and secondary school students) solving non-routine problems to explore their thinking compared to one's own. This included: (i) solving an assigned problem and recording the thought process; observing a peer solve the same problem while thinking aloud, and comparing and discussing thought processes. (ii) selecting a non-routine problem for a secondary school student, solving it and recording thought process; observing a student solving it while thinking aloud and discussing the student's experience with it; comparing processes.

(4) Developing a model for non-routine problem solving, representing it as a flowchart, and applying it to solving a non-routine problem in order to evaluate it.

Stage-3 Activities:

These include participants (i) comparing their post-Stage 2 thinking with their pre-Stage 2 thinking; (ii) comparing their understanding of problems to theory (NCTM, 1989); (iii) comparing their problem-solving models and flowcharts with those from theory (Mason, Burton, & Stacey, 1982; Pólya, 1954; Verschaffel, Greer, & de Corte, 2000); (iv) relating their problem-solving models to an inquiry instructional model for teaching secondary school mathematics; (iv) applying their knowledge to critique a current secondary school mathematics textbook they expect to use; and (v) preparing a lesson plan based on their inquiry instructional model.

Group reflection: Each of the three stages also requires small group and whole-class interactions. This includes participants sharing and comparing their individual reflections in small groups and their small-groups' findings in a whole-class setting.

INVESTIGATION OF THE SELF-STUDY APPROACH

The Self-Study Approach was investigated to determine its effectiveness in helping prospective teachers to develop knowledge of problems, problem solving, problem-solving pedagogy and inquiry-based teaching.

Research Method

The participants were 29 preservice secondary mathematics teachers in the second semester of their two-year post-degree education program. This was their first course in mathematics education, so they had no instruction or theory on problem solving prior to this experience. They also were not taking any other mathematics education course in this semester. The data consisted of copies of all of the participants' written work required for all of the activities. There were also field notes of their small-groups and whole-class discussions. The analysis began with open-ended coding of the data. The researcher and research assistants, working independently, coded the data. Coding included identifying (i) the nature of the participants' initial thinking of problems, problem solving and teaching problem solving; (ii) changes in this thinking, following the Stage-2 and Stage-3 activities; and (iii) the nature of their thinking of inquiry teaching in relation to problem solving. The coded information was summarized for each participant and compared for similarities and differences in their thinking. The findings reported here regarding the effectiveness of the Self-Study Approach are based on common shifts in their thinking. However, there were differences in the nature and depth of these shifts.

Effectiveness of the Self-Study Approach

The approach was effective in expanding and deepening the participants' understanding of problems, problem solving, problem-solving pedagogy and inquiry-based teaching. Their thinking of problems shifted from predominantly routine exercises or word problems to an understanding of characteristics that constitute worthwhile mathematics problems. Their thinking of the problem-solving process shifted from a linear, algorithmic model to an understanding of characteristics that constitute non-algorithmic problem solving. Table 3 is one participant's problem-solving model.

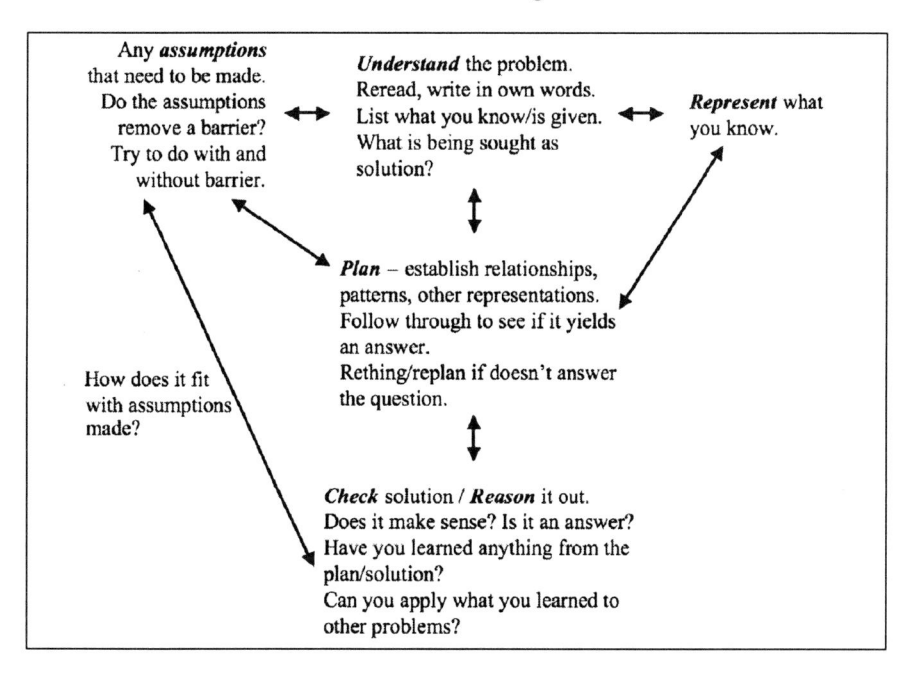

Table 3. Problem-solving model

Any **assumptions** that need to be made. Do the assumptions remove a barrier? Try to do with and without barrier.	**Understand** the problem. Reread, write in own words. List what you know/is given. What is being sought as solution?	**Represent** what you know.
How does it fit with assumptions made?	**Plan** – establish relationships, patterns, other representations. Follow through to see if it yields an answer. Rething/replan if doesn't answer the question.	
	Check solution / **Reason** it out. Does it make sense? Is it an answer? Have you learned anything from the plan/solution? Can you apply what you learned to other problems?	

The theoretical problem-solving models the participants examined in Stage 3 of the approach allowed them to validate, and in some cases refine parts of, their personally developed models based on their experiences, but not to completely change it. Thus, they showed preference for their experiences as a basis of validating what was more meaningful to them.

In terms of teaching problem solving, the thinking of the prospective teachers shifted to a student-centered approach. For example, they explained that they will have students work on a problem first, then share and discuss it in a whole-class discussion; or they will have a whole-class discussion first to understand the problem, have students work in groups or individually, then share and discuss. They will pose questions or prompt the students when they (students) are stuck and use their problem-solving model to help to guide the whole-class discussions.

Finally, the participants were able to create models of inquiry-based teaching by analogy to the problem-solving process and model they developed. They were able to design lesson plans based on these models. Table 4 is one group of participants' inquiry-teaching model.

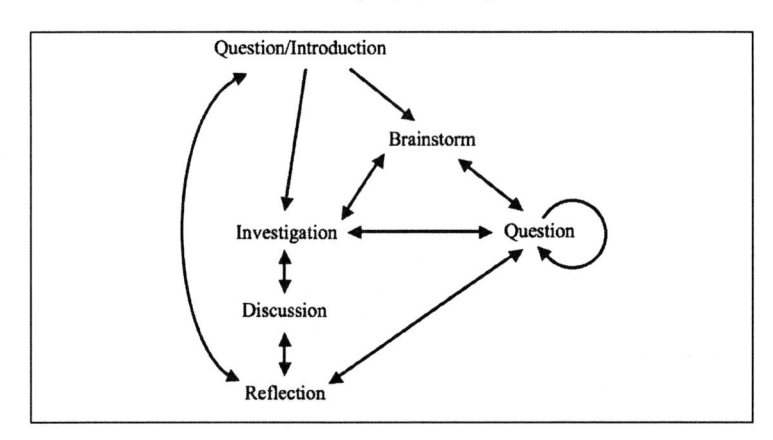

Table 4. Inquiry-teaching model

Overall, the findings suggest that this three-stage approach was effective to facilitate the participants' self-study and construction of useful knowledge about problem solving. However, whether they are able to enact this knowledge in their teaching has not been researched as yet. But holding such knowledge is an important step in getting there.

IMPLICATIONS FOR TEACHER EDUCATION

A self-study perspective could provide an effective basis for an approach to facilitate growth or changes in prospective teachers' understanding and knowledge of problem solving in their teaching. Such an approach should take into consideration all of the following:
– Exploring others as problem solver: e.g., the prospective teacher works with a child or peer to observe, interview, and document information about this child or peer as a problem solver.
– Exploring self as problem solver: i.e., inquiring into one's thinking, learning and instructional practices and developing ability to monitor and to control one's activities when solving problems.

- Exploring the nature/structure of problems.
- Solving problems individually and in small groups without external assistance, e.g., to develop awareness of strategies and skills for solving problems.
- Posing problems.
- Comparing self with others, e.g., peers, students, theorists.
- Formulating an instructional model for problem solving.
- Exploring self as facilitator of problem solving, i.e., to develop an understanding of the teacher's role in facilitating students' problem solving.

The assumption here is that if prospective teachers are to hold knowledge to help them to teach about and through problem solving, they should be provided with experiences not only in solving problems, but also involving the others of these eight factors. It may also be necessary for these factors to be embodied in an integrated experience, as in the Self-Study Approach, and not applied in isolation.

REFERENCES

Chapman, O. (2005). Constructing pedagogical knowledge of problems solving: Preservice mathematics teachers. In H. L. Chick & J. L. Vincent (Eds.), *Proceedings of the 29th meeting of the international group for the psychology of mathematics education* (Vol. 2, pp. 225–232). Melbourne, AU.

Kilpatrick, J., Swafford, J., & Findell, B. (Eds.). (2001). *Adding it up: Helping children learn mathematics*. Washington, DC: National Academy Press.

Lee, H. S. (2005). Facilitating students' problem solving in a technological context: Prospective teachers' learning trajectory. *Journal of Mathematics Teacher Education, 8*(3), 223–254.

Mason, J. (2002). *Researching your own practice: The discipline of noticing*. New York: RoutledgeFalmer.

Mason, J. (1994). Researching from the inside in mathematics education: Locating an I-you relationship. In J. P. Ponte & J. F. Matos (Eds.), *Proceedings of the 18th meeting of the international group for the psychology of mathematics education* (Vol. 1, pp. 176–191). Lisbon, PT.

Mason, J., Burton, L., & Stacey, K. (1982). *Thinking mathematically*. Wokingham, UK: Addison-Wesley.

National Council of Teachers of Mathematics. (2000). *Principles and standards for school mathematics*. Reston, VA: NCTM.

National Council of Teachers of Mathematics. (1989). *Curriculum and evaluation standards for school mathematics*. Reston, VA: NCTM.

Pólya, G. (1954). *How to solve it*. New York: Anchor.

Szydlik, J. E., Szydlik, S. D., & Benson, S. R. (2003). Exploring changes in preservice elementary teachers' mathematics beliefs. *Journal of Mathematics Teacher Education, 6*(3), 253–279.

Taplin, M. (1996). Pre-service teachers' problem solving strategies. In L. Puig & A. Gutiérrez (Eds.), *Proceedings of the 20th meeting of the international group for the psychology of mathematics education* (Vol. 4, pp. 299–306). Valencia, ES.

Verschaffel, L., Greer, B., & de Corte, E. (2000). *Making sense of word problems*. Lisse, NL: Swets and Zietlinger.

Verschaffel, L., De Corte, E., & Borghart, I. (1996). Pre-service teachers' conceptions and beliefs about the role of real-world knowledge in arithmetic word problem solving. In L. Puig & A. Gutiérrez (Eds.), *Proceedings of the 20th meeting of the international group for the psychology of mathematics education* (Vol. 4, pp. 387–394). Valencia, ES.

Olive Chapman
University of Calgary (Canada)

ELKE SÖBBEKE AND HEINZ STEINBRING

6. VISUALISATIONS AS EXAMPLES OF EMPLOYING STUDENTS' POWERS TO GENERALIZE[1]

"Generalization is the heartbeat of mathematics, and appears in many forms. If teachers are unaware of its presence, and are not in the habit of getting students to work at expressing their own generalizations, then mathematical thinking is not taking place." (Mason, 1996, p. 65)

PERSPECTIVES ON THE MATHEMATICS OF EARLY ALGEBRA

If mathematics is understood not as a collection of given rules, procedures and theorems that are transmitted to students by teachers, but as a science of patterns and structures (Devlin, 1997), the principal goals of mathematics education are more about building an understanding of mathematical relations and structures and practicing mathematics on an elementary level solely developing children's calculation skills. For this – and *also* for the successful long-term development of calculation skills – one must also focus on structure, the abstract and the general right from the start of children's mathematics education.

From its beginning, the process of learning mathematics contains situations of change in which a new way of thinking – a new understanding of the previously acquired mathematical concepts and operations and thus a new interpretation of old knowledge – becomes necessary, which is often a demanding task for students. One situation in which new interpretations of mathematical concepts, which are essential for a sustainable understanding of mathematics, take place is the field of fraction calculus and of elementary algebra. On their way from arithmetic to algebra, students have to develop a new awareness for the general, for the variation and the variable. In contrast to elementary arithmetic, in which empirical references are helpful at the beginning (!) in order for children to be able to develop independently their first basic ideas in the field of mathematical operations,[2] the methods and ways of thinking of algebra are not so easily accessible. Algebra is the *lingua franca* of higher mathematics. However, algebra does not gain the meaning and power of such a superior language if its status is restricted to the transformation and calculation of terms, but only if it is used as a "system characterised by indeterminacy of objects, an analytic nature of thinking and symbolic ways of designating objects" (Cooper & Warren, 2008, p. 24).

If the importance of algebra is seen in the way it represented the principles (e.g. commutative principle; balance principle) and structures of mathematics (e.g. field, group and equivalence class) and not in terms of the "behaviours" of algebra (e.g.

simplification and factorisation) which being algorithmic are capable of having their solution processes programmed into calculators or computers (Cooper & Warren, 2008, 24), then it is important for the introduction to algebra to make meaningful learning possible for the students whilst at the same time constructing basic ideas that are sustainable in the long term. Such learning is always situated in the difficult balance between a rather empirical view of concrete objects and actions, of what can be perceived by students and reference to their environment and experienced contexts on the one hand, and a certainly more challenging but in the long run necessary and profitable view on relations and structures. As it seems, there is no lasting way to avoid this difficulty and challenge; if one "protects" the students for too long (acting out of the fear of asking too much of the children, which is widely spread in everyday teaching) from a view onto the *abstract*, onto the *generalisable*, they often lack the intellectual means of handling the system of mathematics, full of relations, structures and patterns, successfully in the long run: "There is a stage in the curriculum when the introduction of algebra may make simple things hard, but not teaching algebra will soon render it impossible to make hard things simple" (Tall & Thomas, 1991, p. 128).

Mason has been trying to enervate these fears by referring to other learning contexts of students, for instance by pointing out that even young children are able to abstract from the concrete and to deal with a general conceptual level intellectually when it comes to linguistic development. "Every learner who starts school has already displayed the power to generalise and abstract from particular cases, and this is the root of algebra. The suggestion made is that expressing generality is entirely natural, pleasurable, and part of human sense-making" (Mason, Graham, & Johnston-Wilder, 2005, p. 2).

LEARNING TO "SEE" STRUCTURES IN ARITHMETIC BEGINNERS' INSTRUCTION

The necessity of focussing on the structures and relations in mathematics, on the variable and the general, however, is not only essential in higher education, for example at the transition from primary to secondary level. The particular character of mathematical knowledge makes it necessary even for *arithmetic beginners' instruction* not to remain restricted to exclusively empirical references to concrete things and situations, but to initiate and to develop a broader and for the children also a new relational view onto elementary numbers and operations.

Mathematics takes on the structural aspects of reality; it processes them into concepts, theories and algorithms. In this regard, mathematics is, from the building of first number and form concepts, self-referential and autonomous. Thus, it is fundamentally different from the sciences, whose concepts – at least partly – refer to empirically accessible reference contexts. "Historical, philosophical and epistemological analyses have been elaborated as a basis of characterising mathematical knowledge ultimately as theoretical knowledge. A central criterion of theoretical mathematical knowledge – also observable in the course of its historical development – lies in the transition from pure object, or substance thinking, to relation or function thinking" (Steinbring, 2008, p. 307). This approach of a new interpretation of mathematical knowledge has to be introduced with children in elementary arithmetic teaching, in order for them to be able to develop successful calculation strategies for elementary arithmetical operations in the following school years. Current research (e.g. Carraher et al., 2006) points out that such methodological tricks, reductions or avoidance strategies are not necessary or helpful even when used for younger students: "research has shown that young students can generalise to abstract representations, and that such activity results in better understanding of mathematics structures in later years" (Morris, 1999). As recent psychological research has pointed out as well, the opposite of this widely spread "avoidance strategy" is helpful and important for the learning of mathematics: "An instruction, which, out of respect for the weaker students, aims at a simple, everyday-life like program and avoids abstract mathematical concepts is amiss. Elsbeth Stern: 'The opposite of good in this case is not bad, but meant in a good way: Especially with weaker children an everyday-life like way of teaching has the least effects, quite on the opposite an abstract program brings the biggest progress. This is because strong students discover mathematical structures in easy exercises by themselves, but the weaker students need specific support and encouragement in order to do so'" (Max-Planck-Gesellschaft, 2005).

MATHEMATICAL MEANS OF VISUALIZATION AS MEDIATOR BETWEEN STRUCTURE AND CHILD

Yet, how is it possible to talk about mathematical concepts and operations with elementary school children if these are something so abstract that ultimately one cannot make them perceivable in their actual meaning in a simple way? Necessary for this are *means of mediation*, which make it possible to represent abstract mathematical concepts as well as to talk about these with children and to enable them to think about them. By means of an active and mental use of means of visualisation the child is to be supported in constructing an adequate mental imagination of these abstract mathematical contents.

But even by using mathematical means of visualisation, one can*not* ultimately spare the child the decisive and necessary step of generalisation and abstraction of the concrete features of the material. Means of visualisation do not *transfer* – according to a widely spread opinion and hope – the mathematical concept to the child; they *mediate* between the mathematical concept and the child. Considering this background, means of visualisation are not only means of help for calculating, but rather their function expands *substantially* by means of a new view on mathematics: If learning mathematics is understood as a process of the children's more and more differentiated way of understanding and interpreting of abstract patterns and structures (cf. Steinbring, 2005), then means of visualisation no longer have a purely didactical or methodological status, but an epistemological one: They are essential means of cognition and of articulation. This function will be clarified by the following example (fig. 1).

This example is not to be centred on the concrete features of the resource, but on the abstract; the relations and the structures within the resource. Thus, what is

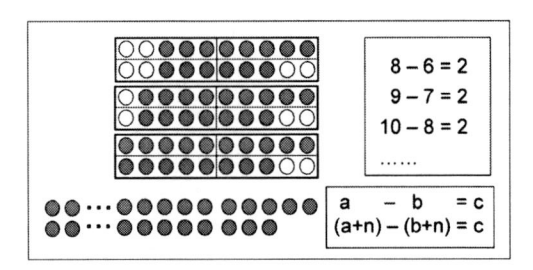

Figure 1. Means of visualization as an epistemological tool.

decisive for mathematical cognition in the figures is not the colours or the number of points; it is rather the *function* that the concrete feature of the resource takes *for* something. The arrangement of the counters in the twenties fields makes the interpretation of a superior relation possible: this is, that the difference of two numbers stays the same when for example both numbers (minuend and subtrahend) are raised by the same number of counters. And only the abstract structure of the representation leads to the insight that this *always* has to be like this. This means, the structure of the representation makes the understanding of a mathematical legality (e.g. as here "the constancy of the difference") possible, but it cannot be read directly or immediately perceived with ones senses; it must be actively interpreted into the representation.

It becomes clear that mathematical means of visualisation can mediate between the mathematical structure and the student's thinking because of their special "double nature" (they are on the one hand *concrete objects*, which can be dealt with, which can be pointed at, which can be manipulatively changed, and at the same time they are *symbolic representatives* of abstract mathematical ideas). Yet, this mediation does not work by itself, by the children working with means of visualisation often enough. It is a new and most likely unusual challenge to take this *new* point of view. Therefore, students have to learn already in elementary school to see the concrete material *also* in this special function as the representative of mathematical structures. The teacher's task is to specifically structure and accompany such a learning process, in which means of visualisation themselves are made a *topic* of mathematics instruction. Dreyfus (1991) has described different stages in mathematical learning processes, which can gain relevance also in regard to this new, just described challenge, and be used for the arrangement of such learning processes: "learning proceeds through four states, namely, using one representation, using more than one representation in parallel, making links between parallel representations, and integrating representations and flexibly moving between them" (cf. Dreyfus, 1991 in Cooper & Warren 2008, p. 3). When relating this to the example described above (fig. 1), the potential of this use of one or several – similar – means of visualisation becomes obvious: At first, the child uses one single representation of the 20s field, perhaps in order to compare "only" two sets (represented by the red and blue counters) with each other and to find out their

VISUALISATIONS AS EXAMPLES

difference. By means of the work with different representations of the 20s field, which is directly offered by the teacher in an operatively structured form, observations (first on the phenomenological stage of concrete material) can be initiated ("8 – 6 = 2", "9 – 7 = 2", "10 – 8 = 2", see fig. 1). An insight into the mathematical legality "constancy of the difference", however, only becomes possible when the child relates the different representations of the 20s field to each other, structurally compares, integrates and flexibly interprets them.

USING VISUALISATION TO DEVELOP STUDENTS' POWER TO GENERALIZE

The following example is to show how elementary school students deal with the difficult challenge of using concrete material on the one hand and at the same time to interpret this material more and more in its function as a representative of relations and structures and thus to focus on the abstract and generalisable in the representation. The scenes presented are not to undergo a thorough methodological analysis (for a careful analysis see Söbbeke, 2005). Rather, the excerpts of the interview are to serve to show the change in Jennifer's interpretations as an example and also to accompany and enable a better understanding of her way from a rather empirical view of the concrete features of the material to a more open and general view onto relations and structures within the diagrams.[3]

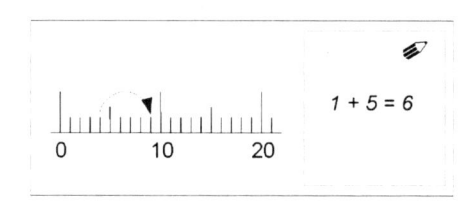

Figure 2. Number line 1.

Jennifer:	Well, here I see the one then. [*moves her finger down to S4*].
Interviewer:	And the plus 5, where do you see that?
Jennifer:	Well [*points at S4*], because that is now a big arc [*moves her finger along the arrow to S9, points at S9*]. Because actually then some think, well one should probably only add the two [*points at S4 and S9*].
Interviewer:	Mhm.
Jennifer:	But that is so, here starts, that goes on here [*points at S 5,6,7,8,9*] up to this arrow, you know [*moves her pen from the top of the arrow to S9*].

At the *beginning* of this interview scene, Jennifer is asked to find an exercise, which fits the representation (fig. 2) "particularly well". She writes down the exercise "1 + 5 = 6" and thus is asked by the interviewer to explain this interpretation. Jennifer's interpretations at the beginning of this interview are characterized by a very empirical way of proceeding, as she does not use the complete representation with its structural relations for the construction of her exercise, but only refers to selected, empirically comprehensible and countable objects of the representation in order to construct an exercise:

> First, 6 lines below the arrow are being counted; then, proceeding from this number the exercise "1 + 5 = 6" is formulated. What is remarkable in this context is that the arrow is not used as a true part of the diagram – in the intended, structure-related way – but as a structure-independent sign with specific requesting character. Against this background the arrow becomes an element of the representation which, largely independently from its position in the number line always gives the same "direction", to distinguish the six lines below it from the other scaling lines.

Subsequent to this, Jennifer receives a *second* number line diagram (fig. 3), for which she writes down the same exercise ("1 + 5 = 6") and names it as very suitable. This way of constructing the exercise corresponds with her first interpretation and reasoning, in which she focussed on the concrete features of the material, meaning that only the set of lines below the arrow need to be counted and proceeding from this, an exercise is constructed.

In the *course* of the interview, after the interviewer has pointed out the caption of the diagram to Jennifer, changes in Jennifer's remarks can be observed. By and by, she becomes aware of the possible structure in the means of visualisation. Thus, Jennifer observed and names the numbers below the scaling lines as "hundreds numbers" and tries to relate these to her previously constructed exercise.

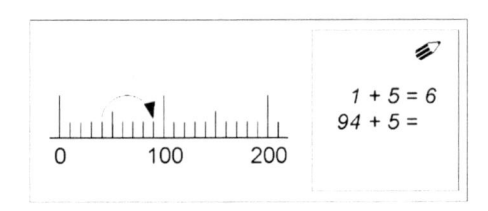

Figure 3. Number line 2.

Jennifer:	That is then the 99, no, mm, the 89 [*waves her hands*], or not? 89? No, a 94.
Interviewer	Mhm. Why?
Jennifer:	Because, ehm, because here I calculated 99, 98, 97, 96, 95, that are then 94 [*with every number she points at one line in the series S9,8,7,6,5,4*]. So in fact I would have to calculate 94 plus 5.

Her interpretations are now characterized by in the interplay of empirical and an initial structurally-oriented way of proceeding, which means that, for the first time, the arrow is integrated into the exercise construction as a structurally-related part of the representation. Altogether, however, these initial structurally oriented interpretations are affected by a recurring orientation towards concrete features and objects (the single lines). For instance, the scaling lines are again interpreted as representatives of countable single steps or as a number line growing by "one" at a time and not in a more abstract sense as representatives of "thought" tens steps.

Finally, Jennifer receives a *third*, now unlabelled number line (fig. 4). After she has written down the exercise "11 + 7 = 18", the interviewer asks whether the scaling line S11 could also represent a different number and makes Jennifer aware of the missing labelling. Towards the interview's *end*, one can see how Jennifer carries out an already very gradual, more flexible and more abstract interpretation of the diagram. While her interpretations and the exercise construction first relate to the number range from 0 to 20 ("11 + 7 = 18") and thus remain with a close reference to the concrete visible objects of the representation (the lines could be counted), she now interprets the visual pattern on the unlabelled number line.

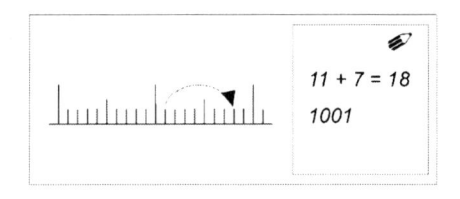

Figure 4. Number line 3.

Interviewer: Mhm. So could this also be a different number now? [*points at S11*],
Jennifer: Mm, mm [*negatively*].
Interviewer: but there is no number written underneath now.
Jennifer: Mm, mm [*negatively*], not like this (...). But that is, that could also be the hundreds row.
Interviewer: Aha.
Jennifer: I don't believe that.
Interviewer: Just try.
Jennifer: Yeah, because that could in fact be <u>anything</u> [*moves her finger along the number line*]. It could, eh, because there only is the 10s row and here the 100s row and maybe also the 1000s row.
Interviewer: Mhm.
Jennifer: Well, this couldn't really be the 1000s row.
Interviewer: Why not?

The new interpretations of the number line diagram could have been affected by the two previous exercises (figs. 2 & 4), which referred to the number ranges from 0 to 20 and from 0 to 200. By means of her several active attempts of interpretation and the intensive examination of the three representations, which have been deliberately offered in this sequence (which can be related to each other as well as being used in a comparing and reflecting way), as well as by means of the interaction between Jennifer and the interviewer, she finally develops a more flexible view of the unlabelled number line, which she interprets in several structurally new ways.

Her interpretations are no longer bound to concrete visible objects; rather the scaling lines for instance can represent different numbers, which cannot be immediately empirically perceived. By Jennifer ultimately interpreting the scaling line as "1001", however, it becomes obvious that the distance between two lines still represent the unity "one". In this regard, there are no extensive new interpretations of the diagram in this phase. Still, Jennifer's remark "that could be anything" shows a first cognition of the relations and structures within the diagram.

SUMMARY

As Mason has shown in many of his research contributions, there is a host of possibilities for exploring and expressing generality (e.g. generality in number patterns, geometrical patterns, diagrams, pictures and picture sequences; cf. Mason, 2005). Lannin (2005) distinguished in this context between generalisations from iconic/visuals and numerical representations, and argued that the iconic is a better representation to lead to process generalisation. As an empirical study (Söbbeke, 2005) has shown, elementary school students are often able to develop and express first ideas of structuring within and with the help of visual representations (cf. the example of Jennifer given above). Because of their theoretical ambiguity and their particular double nature as concrete materials which represent abstract relations and structures, visual representations offer a particular potential in order to support and assist children in their explorations of abstract structures and relations.

This potential, however, can only have an effect if children experience from a particular *teaching and learning culture* that visual representations are more than definite images that help them calculate. Mason (1996) ascertains that the interpretation of exercises and examples, which are offered to the students by the teacher during the classes, are influenced by the education culture, the habits and expectations of the students: "When teachers (...) do an example for students, their experience is often completely different from that of their audience. For the teacher the example is an example of something; a particular case of a more general notion. As the teacher goes through the details the specific numbers or items are experienced as placeholders, as slots in which different particulars could appear. For the student the example is not seen as illustrating a generality but as complete in itself" (Mason, 1996, p. 67). Referring to the use of visual representations as well, children first have to learn not to see these as images with a definite meaning, but to *understand* their variable character. The students' implicit expectations of what is a correct or false interpretation of a visual representation must be extended; for instance by the teacher asking for possible different (also suitable) interpretations. Visual representations do not have one single, definite meaning, but rather the ambiguity they contain in principle makes it possible to see *manifold* relations and structures. Only if teachers do not use visual representations solely as methodological aids, but – in their epistemological function – as means of cognition and of exploring structures, relations and ambiguities together with the children, visualisations can display their power as "Examples of Employing Students' Power to Generalize".

NOTES

[1] The title is in acknowledgment of Mason (2001).
[2] At this point we remark that, even in the first years of elementary mathematics instruction, an exclusively empirically-founded construction of mathematical knowledge is not enough in order to build sufficient sustainable basic ideas for the following higher education (cf. paragraph 2 in this chapter).
[3] In order to follow the pointing activities of the students, the strokes in the number line diagrams are named by S0, S1, S2 and so on:

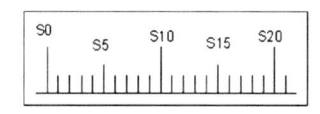

REFERENCES

Carraher, D., Schliemann, A. D., Brizuela, B. M., & Earnest, D. (2006). Arithmetic and algebra in early mathematics education. *Journal of Research in Mathematics Education*, 37(2), 87–115.
Cooper, T. J., & Warren, E. (2008). The effect of different representations on Years 3 to 5 students' ability to generalise. *ZDM Mathematics Education*, 40(1), 23–37.
Devlin, K. (1997). *Mathematics: The science of patterns.* New York: Scientific American Library.
Lannin, J. (2005). Generalization and justification: The challenge of introducing algebraic reasoning through patterning activities. *Mathematical Thinking and Learning*, 7(3), 231–258.
Mason, J. (1996). Expressing generality and roots of algebra. In N. Bednarz, C. Kieran, & L. Lee (Eds.), *Approaches to algebra: Perspectives for research and teaching* (pp. 65–86). Dordrecht, NL: Kluwer.
Mason, J. (2001). Tunja sequences as examples of employing students' powers to generalize. *The Mathematics Teacher*, 94(3), 164–168.
Mason, J., Graham, A., & Johnston-Wilder, S. (2005). *Developing thinking in algebra.* Milton Keynes, UK: The Open University in Association.
Max-Planck-Gesellschaft. (2005). *Presseinformation, 14. April.* Mitteilung zu einer Publikation von Prof. Elsbeth Stern: Zahlenstrahl zündet Geistesblitze, Max-Planck-Institut für Bildungsforschung hat untersucht, wie Kinder Mathematik lernen.
Morris, A. K. (1999). Developing concepts of mathematical structure: pre-arithmetic reasoning versus extended arithmetic reasoning. *Focus on Learning Problems in Mathematics*, 21(1), 44–72.
Söbbeke, E. (2005). *Zur visuellen Strukturierungsfähigkeit von Grundschulkindern -Epistemologische Grundlagen und empirische Fallstudien zu kindlichen Strukturierungsprozessen mathematischer Anschauungsmittel.* Hildesheim: Franzbecker.
Söbbeke, E. (2008). "Sehen" und "Verstehen" im Mathematikunterricht – Zur besonderen Funktion von Anschauungsmitteln für das Mathematiklernen. In É. Vásárhelyi (Ed.), *Beiträge zum Mathematikunterricht 2008.* Münster, DE: Martin Stein Verlag.
Steinbring, H. (2005). *The construction of new mathematical knowledge in classroom interaction – an epistemological perspective.* Berlin, DE: Springer.
Steinbring, H. (2008). Changed views on mathematical knowledge in the course of didactical theory development – independent corpus of scientific knowledge or result of social constructions? *ZDM Mathematics Education*, 40(2), 303–316.
Tall, D., & Thomas, M. (1991). Encouraging versatile thinking in algebra using the computer. *Educational Studies in Mathematics*, 22(2), 125–147.

Elke Söbbeke & Heinz Steinbring
Universität Duisburg-Essen (Germany)

Mathematical
Action &
Structures
Of
Noticing

Part 2:

**THE DISCIPLINE OF NOTICING:
MATHEMATICAL PEDAGOGY AND PEDAGOGIC
MATHEMATICS**

Part 2:

THE DISCIPLINE OF NOTICING: MATHEMATICAL PEDAGOGY AND PEDAGOGIC MATHEMATICS

If mathematical thinking is bound up with attending to features of problems at the same time as sharpening awareness of the processes of thinking mathematically, then the focus of research on teaching and learning and the activity of teaching and learning is that of noticing. At the same time research needs to be systematic, careful, thorough and, essentially, communicated. Thus what is required is a *discipline* of noticing, bringing together what is occurring in the inside, as researchers and/as teachers attend to students and to mathematical activity, with a disciplined process. This has been developed in John's work over the years, and encapsulated in his 2002 book *Researching your own practice: the discipline of noticing*.

Perhaps 'encapsulating' is the wrong word as it conveys an end to the development of a perspective, an approach. Both John and his colleagues and research students have continued to work with and develop the discipline of noticing, and so it remains work in progress. This idea has proved controversial in the international community, with its expectations of verifiable and replicable research studies. The discipline of noticing, and the consequent approach to mathematical pedagogy, has been a stimulus and challenge to the positivist orientations in research in the field, whilst shifting debates around evidence, not a rejection of the possibility of learning from research. What makes a study research, from this perspective that can be accepted by the community is its ability to stir resonance and curiosity in the reader.

This part of the collection begins with a chapter by Dave Hewitt in which he takes up the challenge, offered by Caleb Gattegno, that a curriculum can be conceived of, and devised, in terms of awareness. That awareness is in part what distinguishes mathematics from other subjects but also is a commonality to the whole of mathematics that might enable students to get away from remembering, as the whole task of school mathematics. This is followed by Tommy Dreyfus and John Monaghan who see a powerful overlap between the work of Vasilii Davydov in Soviet Russia, with his emphasis on the ascent from the abstract to the concrete, rather than the other way around, the progressive decontextualisation of mathematics, and John's emphasis on engagement in the process of drawing away from the concrete. Their discussion is based on a decade of their own approach, which they call *Abstraction in Context*.

The third chapter, by Elena Nardi, draws from her extensive work with University mathematicians and is based on the idea, from John and also from Barbara Jaworski, that learning about and developing one's teaching is sustainable and powerful when it is engaged upon by teachers, at any level, with a research orientation. That can often best be undertaken in collaboration with researchers and perhaps with other teachers, though University lecturers are perhaps the most reluctant to recognize the need for good teaching and certainly for collaboration on learning about teaching. The orientation to recognizing, and therefore fostering, the generalisation in thinking mathematically that elementary pupils are already engaged in, is picked up by Marta Molina, Rebecca Ambrose, Encarnación Castro and Enrique Castro in the following chapter. They join in the growing attention to and body of research on early algebra, basing their arguments on John's writing with Barbara Spence on *knowing to act*.

Alf Coles, in the fifth chapter, confronts directly the discipline of noticing with the expectation that research will find out useful things for researchers and teachers. He sets this debate in the context of aesthetics, and says "The aesthetic dimension privileges the personal and transformative; the scientific privileges the social and verifiable." The final chapter, by Laurinda Brown, puts Francisco Varela's thought together with Caleb Gattegno and with John's approach in emphasising spirituality, though not as commonly understood through organized religions. Laurinda offers some activities to assist readers in engaging with her thoughts, as well as some exemplars from her own research with student teachers on developing their awareness.

DAVE HEWITT

7. TOWARDS A CURRICULUM IN TERMS OF AWARENESS

The notion of a curriculum in terms of awareness came from Gattegno (1971) although he never actually articulated the details of what such a curriculum would look like. He mentioned the phrase as part of a list of topics which could be investigated in schools:

> Awareness of relationships per se is what distinguishes mathematical from other thinking. Is it possible to offer a complete mathematics curriculum in terms of awareness? Is it possible to replace the linear presentation of mathematical ideas by a variety of entries into the field, all starting from scratch and each calling for special awarenesses, and have our students reach at least as good a grasp of mathematics as is currently attained by the best learners? (p. 91)

There is much within this quotation, not least a delineation of mathematics from other thinking. In many ways I took up this challenge practically when I was teaching in schools but here I try to articulate what a curriculum in terms of awareness means for me and exemplify this with the topic of fractions. The desire of getting away from a linear progression of mathematical ideas I see as shifting from the idea that in order to know something on the curriculum I need first to have learned a long list of previous items on the curriculum. This can result in the new item either being delayed until I have learned the rest first, or I may be asked to recall previous curriculum items in order to engage with the new item being presented. If I happen not to recall some of these items then the whole lesson falls apart and the teacher begins to re-teach me what I have forgotten and we never quite get to what is new. The whole process can feel like a lot of hard work both for the teacher and for students.

COST

Gattegno (1972) put forward the idea of a Science of Education and began to discuss this in terms of the "investment" from students and the relative "return" on that investment. Following on from Gattegno's work I looked at the notion of economy of personal time and effort in the learning and teaching of mathematics (Hewitt, 1994) with John Mason as my supervisor. How can we reduce the effort required to learn particular things which appear on the mathematics curriculum? Gattegno talked about the cost of learning mathematics and introduced the idea of *ogdens* – units of cost to come to know specific things. He gave an example of learning number names and differentiated between the number of separate words which someone needs to memorise and the processes involved in putting those names together. For example, an *ogden* needs to be paid for memorising each of the words *one, two three four, five, six, seven, eight,* and *nine*. But then with the additional cost of memorising *-ty* and *hundred* numbers such as *five hundred and six-ty nine* come for "free". Through such analysis he worked out that only 18 *ogdens* need to be paid in order to learn all the number names from 1–999 (Gattegno, 1988). The words which have to be memorised I describe as arbitrary (Hewitt, 1999) in the sense that they are socially agreed labels for the numbers which could have been chosen to be something else, and indeed are in other languages. Names and conventions are socially agreed and for a learner they can feel arbitrary choices even if there are historical and linguistic stories behind their choice. Examples are learning names such as *isosceles, commutative* and learning conventions such as the order of Cartesian co-ordinates and that there are 360 degrees in a whole turn. Due to the nature of the arbitrary they have to be memorised by a learner as they cannot be worked out. I differentiate these things from what is *necessary* in mathematics which are those things for which there are reasons for why they must be how they are, given initial assumptions, Examples include the angles in a triangle in Euclidean geometry adding up to half a whole turn, and $2 + 3 = 5$. These things can be worked out and as such I argue that the educational challenge is to educate students' awareness so that they come to know these through their own awareness rather than as things to memorise. In order to reduce the cost for students it is desirable to leave to memory only those things which need to be memorised and this will result in reducing the *ogdens* paid for any learning and instead focus attention on educating awareness.

AWARENESS

Gattegno refused to define awareness, instead he gave examples of how we do already know about awareness or questions to help us become aware that we are already aware of awareness! One such question I found helpful was: "What is the source of my knowing – where is the source of my knowledge which is not information?" (quoted in Berrington Davies, 1986, pp. 24–25). I understand "information" to include the arbitrary, which I have had to memorise anyway. However, I may find myself trying to remember some things which are necessary too. For example, I might catch myself trying to remember whether all triangles tessellate, as if it were a fact which had to be memorised. Later I might stop trying to remember and imagine rotating a triangle about the mid-point of one side to create a parallelogram, and I can picture how parallelograms tessellate. Suddenly I *know* all triangles tessellate without the need to try to remember something. Gattegno famously said "only awareness is educable" (Gattegno, 1973, p. 2) and the challenge for us as teachers is to work on ways in we can educate the awareness of our students.

What does 'educating awareness' mean? This phrase has been one which has provoked much discussion and I have been influenced by the way in which Mason uses thought provoking phrases to work towards practical implications for one's own practice (Mason, 1987 for this specific phrase; Mason, 2002b more generally). It is such a task I am attempting here; to work on a specific example of Gattegno's (1971) phrase of a curriculum in terms of awareness. First I need to consider what educating awareness might mean. I could tell a student something – explain it 'clearly'. What does explaining it clearly mean? Well I suggest it means that a student can use their existing awareness to relate to what a teacher has said and can account for the phenomenon explained by the teacher in terms of underlying mathematics. The latter part is significant otherwise there is a danger of it being turned into "received wisdom" (Hewitt, 1999) whereby students only come to know it as a fact to be memorised rather than known through their own awareness.

How can we know as teachers whether what we say has become known through awareness or is known as received wisdom? We can ask questions which probe underlying mathematics and not just test that something can be "done correctly". After all it is perfectly possible for a student to get right answers whilst not knowing about the mathematics within their work. For example, I was in a Year 8 (12–13 year olds) bottom set where they were finding the area of triangles. One boy had got every question right and I asked him about what he did on one of the questions. Looking at his work he said he multiplied four by nine and divided by two. I asked why he divided by two and he said that he didn't know because he wasn't there yesterday (which was not relevant since this work had only started this very lesson). I asked why he multiplied the numbers together and he said he didn't know that either. Yet to his credit he had managed to work out what was needed to be done was multiply the numbers written on the drawing together and then divide by two. It was "received wisdom" which got him correct answers but his awareness about area had not been educated. Avoiding the possibility of received wisdom is possible by not telling in the first place. Instead Mason (1987) suggested that to educate awareness requires:
– support for positive (non-judgmental) reflection;
– support for noticing moments when they could have acted differently, or wished they had acted differently;
– support for preparing themselves to notice similar possibilities in the future.
(p. 31)

These points raise for me several issues: as a teacher I need to provide an activity in order for there to be something to reflect upon and the choice of that activity is significant as it can determine the richness of what there is to reflect upon; that there are key moments or incidents which are worthy of particular attention and at times I might want to direct students' attention or ask particular questions; and the last of Mason's points helps me realise that educating our awareness now can lead to being a changed person and carrying that with us into the future.

THE COST OF EDUCATING AWARENESS

For Gattegno, use of awareness was "free" and distinct from the energy required to commit something to memory (an *ogden*). However, there is still an issue about what is being called upon from students in order to learn something new whilst at the same time the notion of costing educating awareness does not seem a productive road to go down as there is nothing so clear cut as there is with a word or convention which needs to be memorised. Gattegno's call in the quote at the beginning talks of a variety of entries and this links well with the notion of awareness since we are aware of a very large number of things and one person can call upon some to become aware of something new whilst a different person might call upon others to become aware of the same thing. There are always different ways to come become aware and so there are a variety of ways in which we might approach a topic as teachers. The challenge for me within Gattegno's quote is that of starting from scratch and calling upon special awarenesses. The starting from scratch links in to the notion of "direct access" which I recall Laurinda Brown talking about when we were both teaching in Bristol in the 1980s. The idea of direct access is to find an entry into a topic which requires as little as possible previous knowledge of the curriculum. It does not matter if students cannot recall what they were doing a few weeks ago because they will not need to recall that. Instead what is presented to students requires only their attention and what Gattegno (1970) called powers of the mind, such as the ability to stress some things and ignore others, abstract rules from examples and use of imagery. Alternatively you might want to consider the set of three pairs of powers which Mason (2002a) identifies whilst acknowledging that there are many others: imagining and expressing; specialising and generalising; and conjecturing and reasoning. Both of these frameworks stress the fact that students arrive with impressive powers that are available to be used in their learning. As Mason (ibid.) says, "as a teacher I am faced with the question, 'Am I stimulating my students to use their powers, or am I trying to do the work for them?'" (p. 107). Direct access offers the challenge to try to find entries into a topic which need as little as possible except what everyone brings within them as human beings – powers of the mind – and students will be expected to use those powers to notice and come to know new things, rather than the teacher doing the work for them. So although students will be involved in much work, the work they will be doing is primarily using their powers rather than having to memorise or recall much previous work. In this sense the cost of learning is reduced.

CURRICULUM IN TERMS OF AWARENESS

There was a joke I heard on BBC Radio 4 in May 2004. It went as follows:

A boy comes home and says to his mother:
Boy: *I learnt today that five apples plus two apples is seven apples.*
Mother: *So what is five bananas and two bananas?*
Boy: *I don't know, we haven't done bananas yet.*

There is a sense here of this boy's education being concerned with learning what the teacher has told him and not using his awareness or powers of the mind to apply what he has learnt to different situations. He is stuck with the little he has been told. One key idea behind a curriculum in terms of awareness is the idea of getting a lot from a little – how can a student become more competent with a mathematics topic or to extend the range of possible mathematics which can be worked on using only the awareness already gained? For example, if I know that two and three is five then I know two chairs plus three chairs are five chairs, two people plus three people are five people and two hundred plus three hundred is five hundred, two million plus three million is five million, etc. The mathematics National Curriculum in the UK states that at Key Stage 1 students (5–7 year olds) should "read and write numbers to 20 at first and then to 100 or beyond" and "know addition and subtraction facts to 10 and use these to derive facts with totals to 20" (Department for Education and Employment, 1999, p. 17). This is an example of a curriculum based upon the notion that smaller numbers are easier and so a curriculum concerned with number should be based on smaller numbers before bigger numbers. A curriculum in terms of awareness, however, would realise that 200 + 300 comes from the same awareness as 2 + 3 and so in terms of awareness it might be dealt with before working on such additions as 7 + 8.

Awareness is being educated all the time. We should not think that we as teachers are so important that we need to teach in order for someone to learn. For example, a student might extend their awareness of five subtract three equalling two to the situation of having forty five subtract three giving forty two. They become aware that what they know already can be applied and extended into situations they had not considered before. I will decide to call such events as extending awareness so as to differentiate between that and the notion of "special awarenesses". I do not know what Gattegno meant when he used the term "special awarenesses" and since he died in 1988 I cannot ask him. So I am in a position of putting the awareness that I have into this phrase. So the way in which I will use it is to name the notion of an awareness which is significant for a particular topic in the curriculum and which might not be so obvious for a learner without some teacher activity. For example, my daughter was 5 years old I recall her telling me that you cannot do 3-5. I asked her to start taking five away from three and she counted down: *two, one, zero* and then stopped. Having established that she had not taken away all five, I asked how many she still had to take away. She could tell me two. So I told her that this is what we write down for our answer "take away two" and wrote -2. This seemed to make sense to her and she was excited about being able to answer such questions and we did more of them. After a while I changed my language to "negative two" and she gradually changed hers as well. What I offered was a series of questions and a way to label. She was unlikely to have done that herself as she was stuck with the phrase "cannot do". So this new awareness that such subtractions can be done and how to label answers is an example of how I will use the phrase "special awareness". Following this special awareness she was then able to extend her awareness to many other situations involving negative numbers without the need for my intervention. Of course, she *might* have developed this special awareness without my involvement at all. Some children do so without adult involvement. So I am not claiming that teacher intervention is necessary for students to develop such awarenesses. It is just that for those students they might get there sooner with such intervention and for others they might not have got there at all.

So one aspect I wish to address is the identification of these special awarenesses. This, I believe, is not an exact science since a special awareness will be relative to the awareness a student already has. However, I believe there can still be much gained from such identification as it can highlight where a particular activity or series of questions might need to be planned in order to help students shift their thinking and other times when less intervention is needed and instead space provided for students to continue exploring.

First I would like to clarify a difference between the notion of a special awareness and the notion of a concept as sometimes analysis of a topic can be done in terms of concepts to be learned. My daughter gained an awareness that subtractions such as $3 - 5$ can be done and she learned how to label answers to such subtractions. This says nothing really about any concept she might have or not have of negative numbers. Her ideas of negative numbers will begin to develop but what is special about this awareness is that the gate has now been opened and a new exploration space has appeared. Thus special awarenesses open gates, they say nothing about what might be learned once going through that gate. If I extend the metaphor further, their significance for a curriculum in terms of awareness is that some fields might not be reached without opening these special gates. As well as the identification of such special awarenesses, to keep with the spirit of economy of time and effort and getting a lot for a little, I will be interested in seeing how few such special awarenesses need to be identified for a particular topic.

AN EXAMPLE OF FRACTIONS

Due to limits of word length I will consider work leading up to addition and subtraction of fractions and I am aware that much more could be said than is included here. So I will take this content as a given but consider ways in which this particular content can be viewed in terms of awareness. So I see the curriculum unchanged in terms of what is considered to be desirable for students to learn. What will make this a curriculum in terms of awareness is the approach taken to that curriculum.

Calling fractions into existence – acting upon, notating and naming

With a curriculum in terms of awareness, we never start with an empty slate as everyone arrives into a classroom with existing awareness. So my first question is how does the topic of fractions come into existence? One way is to use what students already know about division of whole numbers and consider the shift from division as a process to division as an object. So 20 divided by 5 can be written $\frac{20}{5}$

and left as such. This can be given the name of a fraction and is met as a division which has not yet been carried out. At the same time this same notation will also stand for whatever the answer to that division is if it were to be carried out. If students can carry out such a division then the statement $\frac{20}{5} = 4$ can be made. An activity of finding other divisions which also equal four can produce a series of equivalent fractions, such as $4 = \frac{20}{5} = \frac{12}{3} = \frac{40}{10} = \frac{4}{1} = \frac{16}{4} = ...$, and a process can begin of students seeing whether they can notice patterns within these fractions and abstract a rule for finding other fractions also equal to four, such as $\frac{?}{6}$ or $\frac{96}{?}$.

A similar development can take place but starting with the idea of comparing one thing with another. This approach, based on ratio and proportion, is one which some have taken to fractions (see, for example, Carraher, 1996), and a similar development as above can be made within that context, although I will choose to stay with division here.

At present this work can be carried out without the need to call $\frac{20}{5}$ anything other than *twenty divided by five*. However, the new objects that are fractions also have special ways of being named. Gradually during such explorations as those indicated above the names can be introduced and so *twenty fifths* can be said. In order to help students notice rules within the way in which these fractions are named I would argue for $\frac{20}{5}$ to be called *twenty* [short gap] *five-ths* (the gap being significant to differentiate it from $\frac{1}{25}$). This can help students notice the generality that fractions are said: *[numerator name]* followed a short gap and then *[denominator name]-"ths"*. I do not suggest that this is stated explicitly as we are concerned with awareness and students who have already learned so much of their first language are perfectly capable of abstracting such a rule given examples. What is not helpful is if the examples on offer use words such as *half, quarters, thirds* which are all exceptions to the more general rule. A principle to assist with students abstracting rules is to offer the general first and worry about the few exceptions later. If a teacher is uncomfortable with using *five-ths* then maybe careful choice of divisions can ensure that more generalised names are used initially, such as *six-ths*.

The shift of naming from *twenty divided by five* to *twenty five-ths* is also a shift in viewing the same notation as a process and as an object respectively, or possibly as question and answer respectively. When we view this as a division we say *twenty divided by five* but when we view this as a fraction we say *twenty* [gap] *five-ths* or *twenty* [gap] *fifths*. The shift in wording can help establish the new topic of fractions through the language signalling when we are viewing this notation as an example of these new fraction objects. The other aspect of establishing the issue of what is a fraction is also working on the degrees of variation (Watson and Mason, 2006) – what is allowed and what is not allowed to be a fraction? One of the activities above can be extended as follows: $\frac{?}{6} \quad \frac{96}{?} \quad \frac{?}{1000} \quad \frac{17}{?}$. Suddenly there is an issue and students can be aware that there is not a whole number at the bottom in the last case. An example of $\frac{10}{?}$ might result in students writing $\frac{10}{2.5}$ or $\frac{10}{2\frac{1}{2}}$.

Is this allowed? Students can enjoy coming up with strange examples and seeing whether they are going to be allowed or not and this helps establish the boundaries of the field (staying with my metaphor) which is fractions.

What does such work mean in terms of awareness? Well, awareness is being used all the time and we cannot make statements about what any individual student might be aware of at any particular point in such work. However, the gate which has been opened to explore this field has been one of notation and naming. Although this can be viewed as being in the realm of the arbitrary what is important is for students to be aware that such things are possible. The notion that I can write a division down in a particular way and also say that this will be the answer to that division too is essentially algebraic in nature. The notation $\frac{20}{5}$ can be a placeholder for the result of that division even if I may not be able to do that division, just as x is a placeholder for a number which is an as yet unknown. Later we may find out the value of x and likewise later we may be able to write $\frac{20}{5} = 4$.

The creation of the topic of fractions has come from a particular action – in this case it is division of whole numbers – and has involved the writing of this in a particular notation and that notation being given a special way of being named. This combination of acting upon, notating and naming has brought the objects of fractions into existence and so is a special awareness which I describe as the *awareness of fraction as object*. One other awareness which is significant is that there are many divisions, and so many fractions, which are equivalent. Here students can work on what is the same about each fraction which is equivalent and can lead to generating more and more equivalent fractions. The *awareness that each fraction has an infinite number of names* is a second special awareness as it opens the gate to knowing what is the same and what is different in the world of fractions.

An alternative way in which fractions can come into existence is through partitioning into equal parts. Again there is an action, this time partitioning, along with a notation and naming which establishes the topic of fractions. Of course, partitioning may have been involved in developing the idea of division anyway and so links can be present with all that I have said already about division. One difference is that this can lead to a more visual way of considering fractions but more significantly it can lead to different awarenesses. Through partitioning, the role of the denominator is that of naming the number of parts with the numerator indicating how many there are of those parts. The role of denominator as naming and numerator as saying how many can be a useful one later on in fraction work and is not so natural an awareness with fractions being introduced as division of whole numbers. This is not to say that this approach is preferential but it highlights the fact that different approaches can bring with them different awarenesses. For example, the notion of partitioning can sometimes bring less clarity for fractions greater than one compared to dividing whole numbers. I will not go into more detail here but *awareness of fraction as object* and *awareness of a fraction having an infinite number of names* remain special awarenesses just as much if the approach has been through partitioning.

Size

What partitioning can bring is a sense of the unit – what is it we are partitioning? A half of this can be smaller than a quarter of that (see Figure 1).

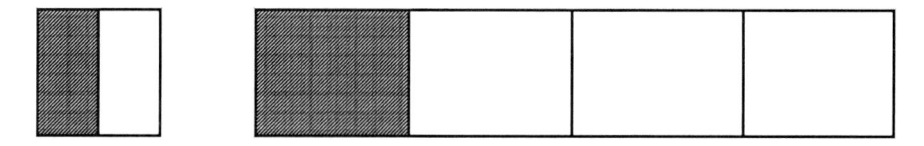

Figure 1. which is bigger, a half or a quarter?

This gives the awareness that size of a fraction is relative to what it is a fraction *of*. Fraction is seen as an operator and with it comes the special *awareness of relative size (fraction as operator)* in that a decision about size will depend upon the relative size of the unit. This can be known through perception initially (as with Figure 1) and later through calculation when division and multiplication are carried out. Such arithmetic work can head towards multiplication and division of fractions although I will not discuss that here.

If we stick with the same unit, or if we have approached fractions as division of whole numbers, then we can find that there are statements we can make about the size of fractions. A half is always bigger than a quarter. Tasks such as partitioning the part of a number line from zero to one, for example, can lead to finding where fractions fit in relation to whole numbers. *Awareness of absolute size (fraction as number)* is another special awareness as it now opens the gate to looking at $\frac{4}{7}$, for example, as not only a placeholder for the answer to four divided by seven but also as a number in its own right. It is no longer just a placeholder for some number I do not know yet, it is a number in itself. This gets students into the world of finding which fraction is bigger or smaller than which other.

Addition and subtraction

Keeping with the idea of getting a lot from a little, if students already are able to carry out some additions of whole numbers (which I will assume they can) then addition of fractions with common denominators comes "free". One of the things which allows us to a statement like *two plus three is five* is the fact that it expresses a generality. It can be:

Two birds plus three birds is five birds
Two tables plus three tables is five tables
Two million plus three million is five million
Two geckos plus three geckos is five geckos
Two x plus three x is five x
Two sevenths plus three sevenths is five sevenths

It really does not matter what we are talking about: *two of something plus three of the same thing is five of those things*. So $\frac{2}{7}+\frac{3}{7}=\frac{5}{7}$ is a result of what we know about two add three. It is not an expression of an awareness about fractions but is an expression of an awareness of addition. In fact I don't even have to know what *sevenths* are. After all, I don't have to know the nocturnal habits of geckos or anything else about geckos to be able to say with certainty that two geckos plus three geckos is five geckos. So, for a curriculum in terms of awareness, there is no need to work on fractions at all in order to know that two sevenths plus three sevenths is five sevenths as this is a statement about addition and not fractions. It is also, of course, just a matter of notation to write this as $\frac{2}{7}+\frac{3}{7}=\frac{5}{7}$. I am not saying that such an addition *should* come before someone knows something about fractions, but just that it could and there is no need for a special awareness – a gate – through which someone needs to pass. I do not know anything about x when I write $2x + 3x = 5x$ (as this is the reason I write x) and likewise I might not know anything about sevenths and still write with certainty $\frac{2}{7}+\frac{3}{7}=\frac{5}{7}$. This approach to adding fractions helps to place the role of numerator and denominator in a linguistic context as was mentioned with partitioning where the denominator names what we are talking about and the numerator tells us how many of those there are.

Addition and indeed subtraction of fractions with common denominators can be developed in the same way as addition and subtraction of integers. Large numbers can be worked with, subtractions resulting in negative answers can be worked with. All will be in the realm of addition and subtraction and it can be useful for students to see that they can be just as good at adding fractions with common denominators as they are with adding integers.

The need to choose common names

Adding fractions with common denominators is one thing but adding any two fractions is another and there needs to be a special awareness that there is a problem if we try to add together fractions with different denominators. What seems important is that students are aware that there is a problem (since such a standard misconception exists with students just adding numerators and adding

TOWARDS A CURRICULUM IN TERMS OF AWARENESS

denominators, for example see Hasemann, 1981). I will offer an activity which is designed to "force" this awareness. Gattegno used the term "force awareness" and my use of the term is that the activity is such that an awareness is likely to occur as a consequence of engaging in the activity. The particular activity I offer below was briefly mentioned by Gattegno within his last talk at the 1988 Association of Teachers of Mathematics conference before his death and here I expand on the original idea.

> Consider that as teacher you have five apples and three pears on a table in front of you. You hold up three apples and ask "what have I got here?" You wait for the reply "three apples". You then keep them in your hand but lower them whilst picking up two apples with the other hand and holding these up. "And what have I got here?" you ask and wait for "two apples". Then you bring the original three apples together with these and ask "what have I got now?" waiting for the answer "five apples".
>
> You repeat the above but with one pear and two pears to get three pears. Then you start again holding up two apples first and then one pear with the other hand (waiting each time until you get "two apples" and "one pear" respectively as replies). Then you put them together and ask "what have I got now?" The key issue here is for students to be aware that they cannot say just *apples* or *pears*. Indeed a new word is needed – *fruit* – which is a common name to both apples and pears.

This activity can raise the awareness that in order to add things together you need to have a common name for them. It is an awareness raising activity – to become aware of when we cannot add and what we need to do in order to be in a position where we can add. The previous special awareness that there are an infinite number of names for fractions allows students to explore whether they can find a common name for each fraction. Once you find that common name then addition is straight forward. This is so whatever things you are trying to add together. It is a required property that things have to have the same name in order to carry out addition and it is no different with the addition of fractions. The awareness that we cannot add things unless they have the same name is an important property of addition and subtraction and can help with a later awareness that $2x + 3y$ needs to stay as it is and cannot be concatenated as is often commonly done by students (Booth, 1984). *Awareness of the need for a common name when adding or subtracting* is a special awareness which is concerned with closing a gate which many students go down and re-directing them along another path. Sometimes closing gates is as important and opening others.

The special awarenesses

I have identified five special awarenesses which I claim are significant gate openings or gate closures which are important to learning a fraction curriculum leading up to the addition and subtraction of fractions. I propose that these are as follows:

Awareness of fraction as object
Awareness of absolute size (fraction as number)
Awareness of relative size (fraction as operator)
Awareness that each fraction has an infinite number of names
Awareness of the need for a common name when adding or subtracting

A curriculum in terms of awareness means that these become the focus of attention for a teacher in planning what activities might be introduced and what questions might be asked for students to become aware of these. Each one of these special awarenesses heralds the beginning of much work where they can come to know and become skilled at what lies within the formal curriculum.

REFERENCES

Berrington Davies, D. (Ed.). (1986). *Working on awareness for teaching and for research on teaching: A seminar by Caleb Gattegno*. Bristol, UK: Abon Language School.
Booth, L. R. (1984). *Algebra: Children's strategies and errors*. Windsor, Berkshire, UK: NFER-Nelson.
Carraher, D. W. (1996). Learning about fractions. In L. P. Steffe, P. Nesher, P. Cobb, G. A. Goldin, & B. Greer (Eds.), *Theories of mathematical learning* (pp. 241–266). Mahwah, NJ: Erlbaum.
Department for Education and Employment. (1999). *Mathematics: The national curriculum for England*. London: DfEE/QCA.
Gattegno, C. (1970). Notes on a new epistemology: Teaching and education. *Mathematics Teaching, 50*, 2–5.
Gattegno, C. (1971). *What we owe children: The subordination of teaching to learning*. London: Routledge and Kegan Paul.
Gattegno, C. (1972). A prelude to the science of education. *Mathematics Teaching, 59*, 34–37.
Gattegno, C. (1973). *An experimental school: A study of a possible renewal of public education*. New York: Educational Solutions.
Gattegno, C. (1988). *The science of education, part 2b: The awareness of mathematization*. New York: Educational Solutions.
Hasemann, K. (1981). On difficulties with fractions. *Educational Studies in Mathematics, 12*(1), 71–87.
Hewitt, D. (1994). *The principle of economy in the teaching and learning of mathematics*. (Unpublished PhD dissertation). Milton Keynes, UK: The Open University.
Hewitt, D. (1999). Arbitrary and necessary, part 1: A way of viewing the mathematics curriculum. *For the Learning of Mathematics, 19*(3), 2–9.
Mason, J. (1987). Only awareness is educable. *Mathematics Teaching, 120*, 30–31.
Mason, J. (2002a). Generalisation and algebra: Exploiting children's powers. In L. Haggarty (Ed.), *Aspects of teaching secondary mathematics* (pp. 105–120). London: RoutledgeFalmer.
Mason, J. (2002b). *Researching your own practice: The discipline of noticing*. London: RoutledgeFalmer.
Watson, A., & Mason, J. (2006). Variation and mathematical structure. *Mathematics Teaching, 194*, 3–5.

Dave Hewitt
University of Birmingham (United Kingdom)

TOMMY DREYFUS AND JOHN MONAGHAN

8. ABSTRACTION BEYOND A DELICATE SHIFT OF ATTENTION

"Men [sic] make their history themselves, only they do so in a given environment, which conditions it, and on the basis of actual relations already existing." (Engels 1894/1968, p. 694)

INTRODUCTION

We have, with colleagues we greatly respect, written about a particular approach to mathematical abstraction during the past ten years. This approach, which we call *Abstraction in Context* (AiC), is inspired by the writings of Vasiliĭ Davydov, a Soviet who was working in this area 50 years ago. AiC is completely at odds with the approach to abstraction as decontextualisation, of stripping ideas from their contextual base and moving from the particular to the general, which dominated Western work on abstraction in the 20th century. We write this chapter because Mason's thoughts on abstraction (Mason, 1989) incorporated ideas akin to Davydov's long before we and others realised their significance. Reflecting on the opening quote from Engels, Mason's achievements are particularly noteworthy because we (in the 21st century) and also Davydov (in Soviet Russia), both with colleagues, developed our ideas in an intellectual environment which valued context; Mason's work in the 1980s in the West was somewhat 'out of synch' with then dominant thoughts on abstraction in mathematics education. The aim of this contribution is to reconsider, twenty years later, the contribution of Mason's paper on abstraction to current thinking and research on processes of mathematical abstraction.

ABSTRACTION AS PROCESS AND PRODUCT

However one views mathematical abstraction, it may be viewed as a process (actions leading up to) or a product (the construct that results from these actions). Mason's title, 'mathematical abstraction as the result of a delicate shift of attention', jumps straight into both – 'abstraction as the result' and 'a delicate shift of attention' as part of the process. Mason's main focus in the paper, however, is on the process of abstraction. On the first page he speaks of "the student's sense of abstract as *removed from* or *divorced from* reality" and states:

S. Lerman and B. Davis (eds.), *Mathematical Action & Structures of Noticing: Studies on John Mason's Contribution to Mathematics Education*, 101–110.
© 2009 Sense Publishers. All rights reserved.

... perhaps this sense of being out of contact arises because there has been little or no participation in the process of abstraction, in the movement of drawing away. Perhaps all the students are aware of is the *having been drawn away* rather than the *drawing* itself. (p. 2)

The *drawing* will vary from person to person depending on who they are (identity), when they are (cultural-historic), where they are (situation) and their personal background (ontogenesis). You, the reader, would not be involved in a process of abstraction if we asked you to engage in a task designed to develop an understanding that different fractional representations of a rational number are possible but a youngster in a well designed environment might be. We, as mature mathematicians, may see members of an equivalence class. It is most unlikely that nine year olds will understand this sophisticated way of expressing what we see but if they think something like *the numbers are different, but they're the same really*, then they have made a great leap, a vertical mathematisation in the RME sense (Freudenthal, 1991), they have constructed a new view, made an abstraction. We believe that this is part of what Mason means when he says "But abstraction is not higher mathematics for the few. It is an integral part of speaking and thinking" (p. 2).

ABSTRACTION AS SHIFT OF ATTENTION

The central idea of Mason's paper is that abstraction consists of a delicate shift of attention. The learner considers the 'same' mathematical object but it appears different to her, which, of course, makes it a different object, at least in the view of this learner. In terms of the example above, the shift may be from a view of $^2/_3$ and $^4/_6$ as two numbers, one expressing that two units are divided into three parts, and the other that four units are divided into six parts, to a view of the part of a unit resulting from each of these divisions and the equality of these parts, hence seeing $^2/_3$ and $^4/_6$ as having the same value and thus being two forms of the same number. The process of abstracting is this shift. Such shifts are central to abstraction and have been central to our and our colleagues' research on abstraction. For example, when analysing a learner's process of abstraction in justifying bifurcations of dynamical systems, Dreyfus and Kidron (2006, p. 216) noted that "Sometimes, one looks at the same object many times and suddenly, it seems different, invested with a new meaning. One does not grasp how one did not see this new meaning previously". This new meaning referred to the learner's shift of attention from a static numerical to a dynamic graphical view of the same mathematical phenomenon and was one of three crucial events in a lengthy process of abstraction. The researchers analysed that process using the AiC framework (Schwarz, Dreyfus & Hershkowitz, 2009). Using the same framework, similar shifts of attention have been identified as central in many other processes of abstraction. Some examples are: shifting from seeing rates of change as a set of values to seeing them as a single varying quantity; shifting from seeing the infinite set $\{1, 4, 7, 10, 13, ...\}$ as a subset of the set of natural numbers $N = \{1, 2, 3, 4, 5, ...\}$ to seeing it as resulting from N by applying the 1–1 correspondence given by $3x - 2$, and thus being of 'the same size' as N; shifting from seeing the sample space of rolling two dice as a minimally structured list of possible outcome pairs to seeing it as a 6 by 6 square array indexed by the outcome on each die. These examples serve to emphasise that mathematical abstractions permeate all of mathematics – an obvious point but, given Mason's focus on algebra, a point worth noting.

AiC has been developed and used for years without reference to Mason's work, and certainly without the explicit intention of discovering or identifying shifts of attention. The shifts of attention appeared as a result of AiC analysis. We are led to conclude that these shifts are inherent in the process of abstraction, and this was the main message of Mason's paper. AiC adds a wider theoretical framework to Mason's description of abstraction as a shift of attention, as well as a methodology to analyse processes of abstraction. Below, we will illuminate some aspects of this theoretical framework and methodology.

SEEING AND NEED

Mason's *delicate shift of attention* proceeds from "seeing an expression *as* an expression of generality, to seeing the expression *as* an object or property" (p. 2). This is very *Masonesque* and is a quite beautiful way of expressing the delicate shift of attention. In this section we briefly attend to *seeing*. Mason is talking about seeing things and seeing things as special things. Why do you see something (in or out of mathematical activity)? Well, sometimes you see something because you cannot avoid seeing it – the big bold arrow pointing to a feature on a graph. But, we claim, you see things (and see them as special things) because you want to see them, when you have a need to see them: seeing a road sign when you are lost which you may overlook if not lost; seeing a function in a table of numbers (these two examples have differences as well as similarity; seeing a function in a table of numbers involves seeing a structure). There are, of course, other conditions on 'seeing' such as prior learning but 'need' is, to us, important and undervalued. 'Need' is essential to our view of abstraction – why abstract if there is no need to abstract? Need in mathematical activity may arise from the demands of the task, interaction with others or the organisation of the artefacts in the activity. Kieran and Drijvers (2007) provide an instance of need arising through a comparison of different techniques:

> ... students were confronting their paper-and-pencil factors with the CAS factors ... certain factoring techniques that they were missing from their repertoire ... they wanted to learn these techniques. This need to understand the factored CAS outputs and to be able to explain them in terms of a certain structure .... (p. 250)

We view need as so important that we place it first the AiC three stage account of the genesis of an abstraction: The need for an abstraction, its emergence, and its consolidation. So we hold that you see "the expression *as* an object or property" because you have a mathematical need to see it as something special in order to get to the essence of what you are studying. Abstraction is not going to happen without there being a need. Mason may have seen this but did not state it.

THE EMERGENCE OF AN ABSTRACTION

Mason speaks of "moving from *manipulating* objects ... to *getting-a-sense-of* ... to *articulating* ... that the process of abstracting in mathematics lies in the momentary movement from articulating to manipulating ..." (p. 3). Yes, in many cases this is a part of the story but there must be something more for how does one 'see' how to get onto the *manipulating/getting-a-sense-of/articulating* cycle in the first place? The need for a new construct is crucial but need is only a prerequisite, it does not explain how to get on this cycle. This is a 'chicken and egg' problem or, as Cole (1996, p. 274) puts it, "how is it possible to acquire a more powerful cognitive structure unless, in some sense, it is already present to begin with"? To our mind Davydov has successfully addressed this problem.

An important aspect of Davydov's work is the idea of 'essential relations', abstract relations which represent a theoretical core insight.

> ... the central or the essential must be detached from the structure of random abstractions, and, furthermore, in thought one must keep to the essence of the matter rather than to accessory mediating qualities that exist everywhere in a complex whole. But where can we obtain a criterion for 'essentialness' – and then how can one be guided by it in choosing *initial* abstractions, for example? In themselves they do not have this criterion. Among them it is impossible to delineate the initial and the subsequent, the central and the noncentral, in an unambiguous way. (Davydov 1990, pp. 277–278)

Grasping the essential relation in a situation is getting the big idea behind it, but this big idea in the emergence of an abstraction is, by necessity, vague and undeveloped. In protocols of learners' work on a task, their grasp of an essential relation is often marked by expressions such as "Aha, I got it" (Ozmantar & Monaghan, 2007, p. 100). This big idea might be the wrong big idea. This does not matter. The next part of the work is to develop the big idea. This requires analysis, establishing relationships between what is known and the big idea; now the learner is on Mason's cycle (or something akin to it). Regarding wrong big ideas – well, mathematics is special amongst subjects in that the development of precise relationships often allows the learner to realise when a big idea is not the right one.

But Davydov leaves the development of an abstraction a bit vague and, from our position as researchers it is difficult to see how to use it in a detailed empirical investigation of learners developing an abstraction. This is where the nested epistemic actions model of abstraction in context comes in. In the next section we outline this model and relate it to Mason's approach.

THE NESTED EPISTEMIC ACTIONS MODEL

To follow the microgenetic development of a student's abstraction, a researcher requires a theoretical framework and a methodological tool to analyse verbal and written data. Abstraction in context provides both. The theoretical framework is based on the idea of vertical mathematisation and Davydov's (1990) dialectic approach where abstraction proceeds from an initial unrefined first form to a final coherent construct in a two-way relationship between the concrete and the abstract. The theoretical framework further adopts Davydov's approach to activity theory to consider processes that are fundamentally cognitive while taking into account the mathematical, historical, social and learning contexts in which these processes occur. In the course of work on a substantial task, learner outcomes of previous activities transform into artefacts in further activity, a feature that is crucial to trace the genesis and the development of abstraction throughout a succession of activities. The kinds of actions that are relevant to abstraction are *epistemic actions* – actions that pertain to the knowing of the participants and that are observable by participants and researchers. The observability is crucial since other participants (teacher or peers) may challenge, share or construct on what is made public. We refer the reader to Schwarz et al. (2009) for a detailed exposition.

The epistemic actions we have found relevant and useful for the purpose are recognising (R), building with (B) and constructing (C), hence the short name 'RBC model'. *Recognising* takes place when the learner recognises that a specific previous knowledge construct is relevant to the problem he or she is dealing with. *Building with* is an action comprising the combination of recognised constructs, in order to achieve a localised goal, such as the actualisation of a strategy or a justification or the solution of a problem. *Constructing* is the central epistemic action of mathematical abstraction and consists of assembling and integrating previous constructs by vertical mathematisation to produce a new construct. It refers to the first time the new construct is expressed by the learner either through verbalisation or through action; in the case of action, the learner may not be fully aware of the new construct. Constructing does not refer to the construct becoming freely and flexibly available to the learner.

These epistemic actions are nested. C-actions rely on R- and B-actions and R- and B-actions make up the C-action. But the C-action is more than the collection of all R- and B-actions: it is what connects between these R- and B-actions and makes them into a single whole unity. It is in this sense that we say that R- and B-actions are nested within the C-action. Moreover, nesting may also include C-actions, which are made for the sake of a more global C-action, meaning that one C- action is nested in a more global one.

Consolidation pertains to the construct becoming freely and flexibly available to the learner. This is not consolidation in the sense of 'drill and kill' but consolidation in the sense of using an abstraction to do further mathematics (Monaghan & Ozmantar, 2006). When a great deal of hard work has resulted in a *delicate shift of attention* it is ... delicate! It behoves us to use it to do something mathematical or it just hangs there, like a work of art, as something to be admired. We see this as implicit in Mason's 1989 paper as the people in the group he reported on take their results to make further explorations. Perhaps it is not explicit because of the diverse but generally advanced level of the people in the group (mathematics undergraduates, graduates and university tutors).

The RBC model constitutes a methodological tool used for realising the ideas of abstraction in context. In this sense, it has a somewhat technical nature that serves to identify learner actions at the micro-level. On the other hand, the model also has a definite theoretical significance; Hershkowitz (2009) discusses the theoretical aspects of the model, its tool aspects, and the relationships between them.

"The shift from looking-at t_n to working with t_n is ever so slight, yet it constitutes a fundamental, and for some people, difficult shift" (Mason, 1989, p. 4). We fully agree and we believe that the manner in which such shifts occur, and the circumstances under which they (fail to) occur merit our close attention. Here we can only speculate on some of the aspects of this process since we do not have data. A person might have a vague idea that t_n is an entity that one can work with because she has similarly worked with other entities, which may have had a simpler structure. Exemplifying, by plugging in some values for n and realising what the notation t_n might stand for and what the role of the n in this notation is, might help that person begin to dialectically move back and forth, in Davydov's sense, between the concrete case and the abstract view of the notation as standing for something concrete might be a step in overcoming this vagueness. We would like to follow this person's epistemic actions. Another person might only work with the numerical sequence without having access to the notation t_n but being able to identify properties of subsequences and other relationships. We would like to follow that person's epistemic actions as well, and learn how each action progressively contributes to building up their views of the sequence t_n until it allows them, in Mason's words, to "work with t_n". The RBC model provides the means for the researcher to do this. Given the relevant data, it allows the researcher to identify the learner's epistemic actions and to follow their micro-history, the way they are referring back to earlier actions and pointing forward to coming ones.

Mason had some of the central ideas later used in the RBC-model: 'looking at' has a taste of recognising, 'working-with' has a taste of building-with, and the shift results from constructing. But Mason does not go into the detail. One reason for this may be that this was not his main concern; another concerns a fundamental difference between constructing and shift of attention which we end this section explaining.

Doing RBC analysis enables (and forces) the researcher to precisely identify the mathematical knowledge under consideration: what are the intended mathematical knowledge elements, what is the construct being built by the learner, to what extent does the learner's construct fit the intended one? Referring to the verb 'abstract', Mason characterises it as "a common, root experience: an extremely brief moment which happens in the twinkling of an eye" (p. 2). Our experience is different. We have observed some sudden constructing actions but we have reported in our research on many students approaching the delicate shift, almost getting the new view and then feeling it escape again; others who clearly see the new view for only a moment before it disappears again and they do not manage to regenerate it: their attention has not shifted, at least not yet. In our conception, constructing may be almost instantaneous but in most cases is not. It can be extended in time, and sudden constructing actions (as well as extended ones) are often difficult for the researcher to identify. Hence, for practical reasons, research studies are most likely to focus on constructing actions of intermediate length (minutes).

ABSTRACTION BEYOND A DELICATE SHIFT OF ATTENTION

Mason never mentioned context, but the shift needs to be made in each context. A slight change of context can easily make the new view disappear. A student can have the new view in one context but be completely unaware of it in a slightly different context.

ABSTRACTION AND GENERALISATION

Mason writes that the "instructions are intended to prompt a move towards expressing a general rule, through experiences of particular cases" (1999, p. 4). This resonates with our Davydov-inspired view of generalisation and abstraction but appears at odds with other views. In this section we consider 'other views' and Davydov's view on generalisation and abstraction before returning to Mason's views.

Jeremy Kilpatrick, in the Introduction to Davydov (1990), wrote:

> Much work on the learning of concepts and principles has assumed that such learning occurs "from the ground up." Students need to see many examples so that they can use induction to form a generalization. The generalization reduces the diversity in the specific examples. Davydov argues that we ought to conceive of learning differently. The specific examples should be seen as carrying the generalizations within them; the generalization process ought to be one of enrichment rather than impoverishment. (pp. *xv–xvi*)

We think Kilpatrick was referring to notables such as Dienes and Skemp and educational psychologists who presented learners with attribute recognition tasks, e.g. tasks which involve sorting different colour-size-thickness shapes into classes. Such tasks bear a resemblance to the work of mathematicians in forming sets based on mathematical attributes: $\{x : \exists\, n \in \mathbb{N}, x = 2n\}$. Such attribute recognition tasks, however, do appear somewhat *impoverished* when compared to the rich tasks that Davydov and Mason (and others) consider. There is a real sense that when you give a learner a task, then you 'get what you give'; the attributes in Mason's task are buried deep and require insight, articulation and connections to be made in the process of abstraction.

One of the points in common between Mason and Davydov is seeing diversity in abstractions and generalisations. We have written above on Davydov and abstraction, but what are his views on generalisation and the relationship between abstraction and generalisation? In generalisation he seeks, as he does with abstraction, the theoretical essence:

> The general relationship that is found by analysis functions as a universal, not because it simply has identical external attributes with its particular manifestations, but because it *is detected* in these particular forms ... the universal relationship that permits determining man's essence ("the production of the tools of his labor") is such because it underlies *all* of the manifestations of human activity .... (pp. 295–296)

Regarding the relationship between abstraction and generalisation he notes:

107

... abstraction and generalization function as two single aspects of the ascent of thought to the concrete. By *abstracting*, man isolates and, in the process of ascent, mentally retains the specific nature of the real relationship of things that determines the formation and integrity of assorted phenomena. In *generalization* he establishes real connections between this isolated relationship and the particular, individual phenomena that arise on its basis. (p. 294)

The relationship between the general and the particular is one that Mason has returned to often over the decades. Davydov provides a theoretical underpinning as to why tasks should "prompt a move towards expressing a general rule through experiences of particular cases" (Mason, 1989, p. 4) and stresses that and why the learner can detect the general, while AiC adds the 'how' to the 'that' and the 'why'. Davydov also distinguishes between abstraction and generalisation as two complementary facets of the same thought process, a distinction that is often glossed over by mathematics educators.

FOREGROUND AND BACKGROUND – THE ROLE OF THE TEACHER

It is a platitude to say that learners need to be offered appropriate tasks in order for abstraction to occur but learners need to be offered opportunities. The important thing is to focus the learner's attention on the intended construct. Within the intentional framework of AiC, a task is likely to be appropriate if, for a specific learner, it is not only motivating but also built so that this specific learner needs a new (to him/her) construct (or alternatively a shift of attention) to achieve the task.

As educators who want to offer students opportunities for abstraction, we should certainly pay careful attention to tasks, the detailed analysis of their content, structure, formulation, and presentation in view of the abstraction potential they offer. Mason's task on sequences could profitably be analysed in this manner. However, the powerful potential for abstraction inherent in Mason's task as presented in the paper stems mainly from the manner it has been implemented. Mason makes use of an observer to focus attention on certain features of the mathematical structure and away from others. When he writes "Thus it behoves the expert to seek ways to draw the students' attention to what is being stressed" (p. 6), he does not exclude the expert who designs the task in such a way as to draw attention to some aspects; but all his examples refer to the tutor or teacher who is assumed to be present at the right moment and help the learner focus. Mason leaves to the teacher a considerable part of the responsibility for initiating the shift of attention and does a wonderful job in showing in his case study how a teacher's hints or rephrasing can create and support the actualisation of such opportunities. The shifts of attention of Mason's learners are delicate as a result of the extremely delicate shifts of emphasis given by his tutor or teacher who acts as Mason's monitor on the shoulder, a "sophisticated mathematical thinker" who is capable of "splitting of attention, one part involved in calculation while another part remains aloof and observant" (p. 4).

Most prominently, the importance of focusing attention is expressed in Mason's insistence on being explicit about switching background and foreground. This is demonstrated most clearly when switching from "sequence with a property" to "property – find appropriate sequences". The movement of the specific sequence into the background, becoming an example, and of the property of the sequence into the foreground, this mutual movement, this switching of foreground and background strikes the core of the process of abstraction; it allows the learner to detect the property in this (and other) examples – as pointed out by Davydov (the last quote above). To what extent students need support from a teacher or tutor in order to enable such a switch of foreground and background is an empirical question; similarly it is an empirical question in what contexts and under what circumstances a teacher or tutor has the power to promote such a shift. But there is no doubt that we, as researchers interested in abstraction, have so far neglected the role of the teacher in processes of abstraction and that we can learn a great deal from Mason's abstraction work with teachers.

CONCLUSION

Mathematics education as a scientific undertaking has a tendency to produce a large number of theoretical ideas and frameworks that are each being used by a rather small group of educators and researchers and tend to remain disconnected from each other in spite of deep underlying connections. Mason has contributed many theoretical ideas like this. We hope that with this brief contribution, we have added to the efforts to establish links between different theoretical ideas in general (Prediger, Arzarello, Bosch, & Lenfant, 2008) and between Mason's ideas on abstraction as a delicate shift of attention and those of AiC, in particular.

REFERENCES

Cole, M. (1996). *Cultural psychology: A once and future discipline.* Cambridge, MA: Harvard University Press.

Davydov, V. V. (1990). *Types of generalisation in instruction: Logical and psychological problems in the structuring of school curricula* (Soviet studies in mathematics education, Vol. 2, J. Kilpatrick, Ed., J. Teller, Trans.). Reston, VA: National Council of Teachers of Mathematics. (Original work published 1972).

Dreyfus, T., & Kidron, I. (2006). Interacting parallel constructions: A solitary learner and the bifurcation diagram. *Recherches en didactique des mathématiques, 26,* 295–336.

Engels, F. (1894/1968). Letter to W. Borgius. In *Karl Marx and Frederick Engels: Selected works in one volume.* Moscow: Progress Publishers.

Freudenthal, H. (1991). *Revisiting mathematics education.* Dordrecht, NL: Kluwer.

Hershkowitz, R. (2009). Contour lines between a model as a theoretical framework and the same model as methodological tool. In B. B. Schwarz, T. Dreyfus, & R. Hershkowitz (Eds.), *Transformation of knowledge through classroom interaction* (pp. 273–280). London: Routledge.

Kieran, C., & Drijvers, P. (2006). The co-emergence of machine techniques, paper-and-pencil techniques, and theoretical reflection: A study of CAS use in secondary school algebra. *International Journal of Computers for Mathematical Learning, 11*(2), 205–263.

Mason, J. (1989). Mathematical abstraction as the result of a delicate shift of attention. *For the Learning of Mathematics, 9*(2), 2–8.

Monaghan, J., & Ozmantar, M. F. (2006). Abstraction and consolidation. *Educational Studies in Mathematics, 62*(3), 233–258.

Ozmantar, M. F., & Monaghan, J. (2007). A dialectical approach to the formation of mathematical abstractions. *Mathematics Education Research Journal, 19*(2), 89–112.

Prediger, S., Arzarello, F., Bosch, M., & Lenfant, A. (Eds.). (2008). Comparing, combining, coordinating – networking strategies for connecting theoretical approaches. Special Issue of *Zentralblatt für Didaktik der Mathematik – The International Journal on Mathematics Education, 40*(2).

Schwarz, B. B., Dreyfus, T., & Hershkowitz, R. (2009). The nested epistemic actions model for abstraction in context. In B. B. Schwarz, T. Dreyfus, & R. Hershkowitz (Eds.), *Transformation of knowledge through classroom interaction* (pp. 11–41). London: Routledge.

Tommy Dreyfus
Tel Aviv University (Israel)

John Monaghan
University of Leeds (United Kingdom)

ELENA NARDI

9. GAINING INSIGHT INTO TEACHING AND LEARNING MATHEMATICS AT UNIVERSITY LEVEL THROUGH MASON'S *INNER RESEARCH*

INTRODUCTION

In recent years developments in the ways in which mathematics education research is conducted – for example, *Developmental Research* (van den Akker, 1999), *Co-Learning Partnerships* (Jaworski, 2003) – have paved the way for the emergence of research that is collaborative, context-specific and data-grounded and, through being non-prescriptive and non-deficit, it aims to address the often difficult relationship between mathematics education practitioners and researchers. Mason's idea of *Inner Research* (1998) relates deeply to the principles that underlie this type of research. One of these principles – a fundamental one – is that development in the practice of mathematics teaching is manageable, and sustainable, if driven and owned by the practitioners who are expected to implement it (p. 375).

According to Mason the two most important products of mathematics education research are: the transformation in the being of the researcher (p. 369); and the provision of stimuli to others to test out conjectures for themselves in their own context (p. 369). Researching from the inside can be as systematic as more traditional formats (p. 364), such as classical formats of qualitative research. In this sense *Inner Research* is not novel to mathematics education research – its origins lie with Schön's (1987) *Reflective Practice*, bar the occasional use of this framework by teacher education programmes in an overly prescriptive way. *But*, while not entirely novel, it is increasingly present, partly because external research has generally failed to generate effective and enduring reform of practice.

With regard to establishing the validity of *Inner Research* as Mason highlights – in the now proverbial *there are no theorems in mathematics education* (in the spirit of which (Mason, 1998) is largely written) – in education, unlike what is mostly the case in the natural sciences, we need to seek ways of making assertions that are locally valid, make sense of past experience (fit), are in the form of conjectures (only), are testable to a wider population and situations, and are as precisely stated as possible. The links with *Action Research* are obvious in Mason's account – but he does warn about the tendency of *Action Research* to turn into outsider research (p. 362). *Inner research* instead calls for the development of sensitivity (*Noticing*) and awareness (p. 366) and is less interested in approaching educational phenomena from a cause-and-effect angle.

With regard to the outputs of *Inner Research* there is a parallel between Mason's description and Chevallard's *transposition didactique* (1985). Whereas

S. Lerman and B. Davis (eds.), Mathematical Action & Structures of Noticing: Studies on John Mason's Contribution to Mathematics Education, 111–120.
© 2009 Sense Publishers. All rights reserved.

traditional research outputs come to be through a process that (for example) transforms data into typically alienating (to the practitioner) formats, *Inner Research* aims to transform pedagogical insight gained in the process of the research into learnable and teachable insight through conveying a sense of a particular experience, e.g. of a student learning mathematics, in lively and authentic ways. It is mostly for this reason that I think that it makes perfect sense to define *Inner Research* in terms of *research as working on being* (p. 372).

My aim in this paper is to demonstrate the potent nature of Mason's *Inner Research* (1998) in the particular context of teaching and learning mathematics at university level through drawing on a study that was designed and carried out in the *Inner Research* spirit. The study – see the Note at the end of the paper for a brief introduction – invited university mathematicians to engage with reflection on their students' learning and on their pedagogical practice. In what follows I hope to achieve the above aim through: exemplifying the insights offered by the participating mathematicians into their students' learning and their own pedagogical and mathematical practice; and discussing the mathematicians' evaluative comments on the experience of participating in this type of research.

The issue that my exemplification revolves around is *the role of visualisation* in a mathematical argument, an issue that in recent years has been attracting increasing attention both within mathematics and mathematics education (Presmeg, 2006). My exemplification draws on the analyses that led me (Nardi, 2007) to discuss the interviewed mathematicians' perspectives on *students' attitudes towards visualisation*, the *role of visualisation* in student learning and the pedagogical role of the mathematician in fostering a flexible *attitude towards visualisation*. I also discuss the participating mathematicians' *evaluation of participating in the study*. Unless otherwise stated page numbers refer to pages in (Nardi, 2007), mostly the episodes in pp. 139–150, 195–199 and 237–247.

EXAMPLE I: BENEFITING FROM NOTICING – MATHEMATICIANS' PEDAGOGICAL INSIGHT INTO THE ROLE OF VISUALISATION

Students often have a turbulent relationship with visual means of mathematical expression (Presmeg, 2006): when facing difficulty with connecting different representations (for instance, formal definitions and visual representations), they often abandon visual representations – that tend to be personal and idiosyncratic – for ones they perceive as mathematically acceptable. Below I sketch some of the interviewed mathematicians' perspectives on students' attitudes towards visualisation and on the ways in which these attitudes – and ensuing behaviour – can be influenced by teaching. The premise for the discussion is the following mathematical problem (typically given to Year 1 mathematics undergraduates in a Semester 1 Calculus / Analysis course):

> Write down a careful proof of the following useful lemma sketched in the lectures. If $\{b_n\}$ is a positive sequence (for each n, $b_n > 0$) that converges to a number $s > 0$, then the sequence is bounded away from 0: there exists a number $r > 0$ such that $b_n > r$ for all n. (Hint on how to start: Since $s > 0$, you might take $\frac{1}{2}s = \epsilon > 0$ in the definition of convergence.)

One acceptable approach to this is described in the notes below (written by the lecturer of the course the problem originally comes from):

> Let $\epsilon = \frac{s}{2}$ in the definition of convergence. Then there is an N such that $n > N \Rightarrow |b_n - s| < \frac{s}{2} \Rightarrow b_n \geq \frac{s}{2}$. Then, for any n, $b_n \geq r = \min\{b_1, \ldots, b_N, \frac{s}{2}\}$ which is the minimum of finitely many positive quantities, hence is positive.

Questions / issues touched in the discussion included: what responses would you expect from the students to this problem; what difficulties may they face; if you were to discuss this problem with a student how would you do so?

One of the issues that emerged in the course of discussing this problem concerns the fact that, in the second line of the lecturer's notes, a perfectly acceptable part of the argument is to 'leave out' of the inequality the terms b_1, b_2, ... b_N. Why this is helpful can also be visible in a simple picture that portrays the 'boundedness away from zero' of the significant majority of the sequence's terms. Students treated the contingency of such a picture variably. See Table 1 which shows three typical Year 1 student responses and the comments made on them – with regard to the presence, absence and quality of such a picture in the students' scripts – by mathematicians who are participants in the study and who teach these students.

Are the students' responses, or their teachers' comments, surprising or substantially different to what you had in mind? What do these responses and comments suggest about the difficulties these and other students may have with this or similar problems? What feedback would you give these students? How would you help them overcome these difficulties in this case? ...or more in a more general but related case?

First and foremost the interviewees described 'pictures' as efficient carriers of meaning:

> *[talking about* || *as distance]* What the students really need to be thinking about is what || means on the number line and as a distance. But they so often get stuck to the algorithmic habit of *solving* this without knowing what it means. And that stubbornness can be a nightmare.
>
> What I mean by *what it means* is, for example, seeing, what an equality or inequality involving $|x - 1|$ means pictorially on the real line. Once you have seen it on the line, the answer to your question is obvious. That is why I am a huge fan of them using all sorts of visual representation: because the ones who do, almost invariably are the ones who end up writing down proper proofs. (p. 238)

Appreciating this efficiency however is often hindered by students' ambivalence about whether 'pictures' are 'mathematics':

> Students often mistrust pictures as *not mathematics* – they see mathematics as being about writing down long sequences of symbols, not drawing pictures – and they also seem to have developed limited geometric intuition perhaps

Table 1. Three typical student responses commented by the students' teachers (Examples and comments are from p. 140 and pp. 195–199 in Nardi, 2007)

Student N has not left out of his argument a small but significant number of terms in the sequence he is working on. 'Had the student drawn a picture, he would have [this]'.	*(handwritten mathematical work)*

Student N, no picture

Student H, emulates 'the type of picture drawing seen in lectures'. She however 'needed a more helpful picture'. It is encouraging though that both Students N and H pinned down an understanding of ‖ as a 'distance between things'.	*(handwritten mathematical work with diagram)*

Student H, unhelpful picture

Student E has not 'used this diagram as a source of inspiration for answering the question'. Instead 'she drew this, on cue from recommendations that are probably on frequent offer during the lectures, and then returned to the symbol mode unaffected'. So 'there is no real connection between the picture and the writing'.	*(handwritten mathematical work with number line diagram)*

Student E, not benefiting from picture

since their school years. I assume that, because intuition is very difficult to examine in a written paper, in a way it is written out of the teaching experience, sadly. And, by implication, out of the students' experience. It is stupefying sometimes to see their numb response to requests such as imagining facts about lines in space or what certain equations in Complex Analysis mean as loci on the plane. (p. 139)

Evidence of this ambivalence can be found in the students' responses to the mathematical problem cited earlier: consulting a visual representation is described by the interviewed mathematicians as potentially beneficial but the students have made very variable use of these benefits.

Across the interviews the participants' insights into the role of visualisation in student learning revolved around the following four axes:

- Usefulness of visual representations: firm and unequivocal ('Graphs are good ways to communicate mathematical thought', p. 143)
- The usefulness of educational technology, e.g. graphic calculators: caution and concern ('Calculators are nothing more than a useful source of quick illustrations', p. 143)
- Students' varying degrees of reliance on graphs (both in terms of frequency and quality)
- The potentially creative fuzziness of the 'didactical contract' at university level with regard to the role of visualisation

Particularly with regard to the last two, in a nutshell, the discussion of the mathematicians' perspectives on the role of visualisation in student learning is in the light of how the mathematicians employ visualisation in their own mathematical practice. The emergent perspective from the discussion is of the need for a clear didactical contract (Brousseau, 1997), in which students are encouraged to emulate the flexible ways in which mathematicians to-and-fro between analytical rigour and often visually-based intuition (for example, the students are allowed to use hitherto unproven facts, on the proviso that, eventually, at a later stage they will establish these facts formally). Specifically, the discussion concludes with two pertinent observations on the role of visualisation:

- The strongly personal nature of the 'pictures' at work in mathematical thinking:

  ...students end up believing that they need to belong exclusively to one of the two camps, the informal or the formal, and they do not understand that they need to learn how to move comfortably between them. Because in fact this is how mathematicians work! I still remember acutely my own teachers' explanations of some Group Theory concepts via their very own, very personal pictures. I am a total believer in the Aristotelian *no soul thinks without mental images.* In our teaching we ought to communicate this aspect of our thinking and inculcate it in the students. Bring these pictures, these informal toolboxes to the overt conscious, make students aware of them and help them build their own.

And I cannot stress the last point strongly enough: we need to maintain that these pictures are of a strictly personal nature and that students should develop their own. All I can do is describe vividly and precisely my own pictures and, in turn, you pick and mix and accommodate them according to your own needs. (p. 237)

— The potently creative 'fuzziness' of the to-ing and fro-ing between analytical rigour and intuition

Lest we forget some very clever people regarded [for example] the Intermediate Value Theorem not needing a proof either! People like Newton. [...] there is an irony in the fact that validating the truth of the statement in IVT means that all the pictures that students have been drawing are retrospectively true – like drawing the solutions of an equation. This irony in fact is nothing other than another piece of evidence of a constant tension within pure mathematics: that you want to use these methods and occasionally you need a theory to come along and make them valid. And you need these means, diagrams etc., so badly. Yes, they are not proofs but they do help students acquire first impressions, start inventing some suitable notation. (p. 238)

Finally, below are examples of three mathematical topics where the above attitudes towards visualisation are described by the interviewees as useful. The first concerns a geometric approach to problems in the complex plane:

... there are things that are very concrete like geometric problems in the complex plane that I frankly cannot think of a better way of presenting other than through a diagram, for instance to see that $|z - a| = 1$ is a disk. (p. 238)

The other two demonstrate the to-ing and fro-ing between analytical rigour and visually-based intuition in the context of strengthening students' understanding of functional properties through the construction and examination of function graphs.
For example, exponentials:

... let's think about the so called *bigness* of exponentials. This is just about the time that students ought to start realizing how faster exponentials grow in comparison to powers. Through a picture like this they can start seeing this and then, of course, prove it by induction or whatever. A more formal grasp of exponentials comes subsequently when Taylor expansions and Riemann integration are introduced. Then exponentials are seen as inverses of logarithms. These more formal images are not necessarily substantially helped by early pictures but for a primary understanding of exponentials this picture is good enough. And necessary. I don't want to wait until the Riemann integral is formally introduced for the students to have an image of what the graph of $x^{-1}e^x$ looks like. This is a totally unnecessary and unproductive delay. Of course exponentials are also buttons on the calculator and in that sense the pointwise view comes to the rescue – at least temporarily. (p. 240)

And powers:

> ... overall the exercise of drawing the graph of $x^{-1}e^x$ is probably wonderful and very good for them despite these big conceptual holes. And you can get students to generalise x^n for any integer n which also has this nice property of getting bigger as x gets bigger. You can explore how fast this happens. And every now and then you get a surprising insightful comment from a student such as *the degree is not high enough* for our function to do this or that. And that is impressive and reassuring. That they can get to see the dominant term in a polynomial when you are trying to produce its graph. And in fact see these things beyond the range of your graph paper. Use their imagination on how this thing would go beyond what fits in your piece of paper. (p. 241)

EXAMPLE II: MATHEMATICIANS' INSIGHTS INTO THE BENEFITS OF ENGAGING WITH *INNER RESEARCH*

The interviewees juxtaposed, often emphatically, the 'indecipherability', futility and irrelevance of mathematics education research texts they had encountered previously with the potent experience of participating in these example-rich, experience-based interviews (see brief description of the *Student Data Samples* in the study in Nardi, 2007):

> ... it is in these discussions exactly that these sessions have proved enormously valuable already. There are things I will teach differently. There are things that I feel like I understand better of mathematics students than I did before. And I appreciate the questioning aspects of the discussion and I realise how one should be liaising with the other lecturers simultaneously lecturing the students and discussing what things we are doing that confuse them. (p. 260)

The benefits described in these evaluative comments are as follows:

— realising the extent of student difficulty

> ...these discussions are already beginning to influence the way I think about my teaching. I think discussing the examples is a very good starting point, and a well-structured one. By seeing these often terrifying pieces of writing I am faced with the harsh reality of the extent of the students' difficulties. Too often I see colleagues who are in denial and opportunities like this are poignant reality checks! [...] I am therefore grateful for this opportunity to face the music, so to speak.' (p. 261)

— the potential of change through engagement

> There is substance in this; it is important. Suppose you have a schoolteacher. So, here is someone who has to run classes and for some reason or another their view of mathematics is no other than an instrumental one: you apply this

rule, you put this in and you get this out. Suppose that such a person one day meets *Concept Image* and all that. All of a sudden he learns that these things are all out there and that changes that person's professional view entirely. It can change the whole classroom, it can change the whole mathematical process. That is precisely what we want.

A lot of the problems you have to deal with when you meet our students is that they have a very singular view of mathematics, a rather poor view of mathematics. So, I mean, that sort of debate that is happening here is on some of the building blocks around which, it seems to me, if made available at the school level for practitioners, would be hugely interesting. To get away from this sort of mathematics which is quite poor in a way. (p. 262)

— gaining in openness and awareness

I think now I don't have any more answers than when I started <u>but</u> certainly I don't take things for granted anymore, from colleagues or from students.

I think I am much more open-minded on what might be going on inside other people brains. The material that you have got here has given the evidence that sure, it is fascinating glancing in other people's heads.

And I have become much more conscious about the spoken word. What I say can have an impact, saying the right thing at the right time when you get one opportunity to introduce the students for the first time to how mathematics works and not fluff the line. That I think has made a big influence on the way I lecture. (p. 263)

THE SYNERGY BETWEEN TWO COMMUNITIES, BETWEEN MATHEMATICS EDUCATION THEORY AND PRACTICE

Let me use one of the comments made by interviewees as the premise for drawing a parallel with similar comments made in the educational literature, and in particular with some of Mason's *Tactics* (2002). The comment concerns the growing realisation in the course of the study of the need for 'detailed responses' to students' written work:

… examining these pieces of data was something of a reminder, if not a revelation, of the devastating importance of detailed responses to written work. In some sense every not totally perfect piece of written work has an interesting important story to tell that needs to be engaged with and responded to. (p. 263)

The parallel is with Mason's comments (2002, chap. 5) on providing feedback to students. There Mason elaborates on the Tactics: Focusing on what is mathematical; Developing a language; Finding something positive to say; Selecting what to mark; Summarising your observations; and, Providing a list of common errors or a 'corrected' sample of student argument.

The above commonality of perspective, focus of attention and sensitivity is to me a fraction of the potential lying in the synergy between the communities of mathematics and mathematics education – that this and a growing number of studies in the field aim to foster. Or in Michèle Artigue's (1998) words:

> ... we, mathematicians as well as didacticians [...] have to act energetically in order to create the positive synergy between our respective competences which is necessary for a real improvement of mathematical education, both at secondary and at tertiary levels. Obviously such a positive synergy is not easy to create and is strongly dependent on the quality of the relationship between mathematicians and didacticians. (pp. 482–483)

HOW MATHEMATICS EDUCATION RESEARCH IN THE SPIRIT OF *INNER RESEARCH* PROMOTES THIS SYNERGY

I hope that the work I draw on in the above demonstrates that researching from the inside *systematically* (Mason, 1998, p. 364) is feasible: apart from the systematic way in which the *Student Data Samples* were composed (the student examples were selected as typical from large pools of data collected in the course of previous studies; their themes were determined by analyses in these previous studies and informed by our consultation of relevant research), every other aspect of the research attended to the classical 'rules' of qualitative research (for example in the ways in which participants were selected, the focused interviews conducted and the data processed and analysed).

Furthermore, and returning to the idea of *research as working on being*, described in the opening pages of this paper, in my eyes, the *Student Data Samples*, the way of conducting and analysing the interviews with the mathematicians and the dialogic format (in Nardi, 2007) demonstrate a case of Mason's transformation (of the mathematicians and the mathematics education researchers involved) from within. The overall approach to the research exemplified here is a case of the – often ambitious – aspiration of *Inner Research* to convey an authentic sense of a particular experience. Here this experience is the ways in which mathematicians perceive, enact and narrate their pedagogical role. The *Student Data Samples* seem to propel them towards *Noticing* and *'spection'* (Mason, 1998, p. 372) of all sorts and to elicit evidence of, as well as heighten, their awareness. The researchers' questioning of the mathematicians creates space for / propels them towards a richer *mathematical being* (p. 374) – one with an overt, conscious pedagogical dimension, a *pedagogic being* (p. 374). Of particular significance to the researcher in mathematics education is witnessing, in the process, the emergence and demonstration of the mathematicians' pedagogical awareness. And the researcher's *ability to bear* (p. 373) is essential in this – a quintessential element in mastering the craft of a balanced practitioner / researcher relationship.

NOTE

The evidence base of the study on which this chapter draws consists of focused group interviews with mathematicians of varying experience and backgrounds from across the UK. Its analyses, carried out through the narrative method of *re-storying* (Clandinin & Connelly, 2000) have been presented in Nardi (2007), in the format of a dialogue between two fictional, yet entirely data-grounded characters, M, mathematician and RME, researcher in mathematics education. The study is the latest in a series (1992–2004) that: explored Year 1 and 2 learning (tutorial observations, interviews and written work); and, engaged lecturers in reflection upon learning issues and pedagogical practice (individual and focused group interviews). The focused group interviews conducted in the course of the latest of these studies lasted about half-a-day. Discussion in the interviews was triggered by *Student Data Samples* based on the findings of the previous studies. These were samples of students' written work, interview transcripts and observation protocols collected during (overall typical in the UK) Year 1 introductory courses in Analysis/Calculus, Linear Algebra and Group Theory. The findings were arranged according to the following themes: *student learning* (mathematical reasoning, in particular conceptualisation of the necessity for proof and their enactment of various proving techniques; mathematical expression and attempts to mediate meaning through words, symbols and diagrams; encounter with fundamental concepts of advanced mathematics – Functions, across Analysis, Linear Algebra and Group Theory, and Limits): *university-level mathematics pedagogy; the often fragile relationship between the communities of mathematics and mathematics education and conditions for collaboration.*

REFERENCES

Artigue, M. (1998). Research in mathematics education through the eyes of mathematicians. In A. Sierpinska & J. Kilpatrick (Eds.), *Mathematics education as a research domain: A search for identity* (pp. 477–490). Dordrecht, NL: Kluwer.

Brousseau, G. (1997). *Theory of didactical situations in mathematics.* Dordrecht, NL: Kluwer.

Chevallard, Y. (1985). *La transposition didactique.* Grenoble, FR: La Pensée Sauvage.

Clandinin, D. J., & Connelly, F. M. (2000). *Narrative inquiry: Experience in story in qualitative research.* San Francisco, CA: Jossey-Bass.

Jaworski, B. (2003). Research practice into/influencing mathematics teaching and learning development: Towards a theoretical framework based on co-learning partnerships. *Educational Studies in Mathematics, 54*(2–3), 249–282.

Mason, J. (1998). Researching from the inside in mathematics education. In A. Sierpinska & J. Kilpatrick (Eds.), *Mathematics education as a research domain: A search for identity* (pp. 357–377). Dordrecht, NL: Kluwer.

Mason, J. (2002). *Mathematics teaching practice. A guide for university and college lecturers.* Chichester, UK: Horwood Publishing.

Nardi, E. (2007). *Amongst mathematicians: Teaching and learning mathematics at university level.* New York: Springer.

Presmeg, N. C. (2006). Research on visualization in learning and teaching mathematics: Emergence from psychology. In A. Gutierrez & P. Boero (Eds.), *Handbook of research on the psychology of mathematics education* (pp. 205–235). Dordrecht, NL: Sense Publishers.

Schön, D. (1987). *Educating the reflective practitioner.* San Francisco, CA: Jossey-Bass.

van den Akker, J. (1999). Principles and methods of developmental research. In J. van den Akker, R. Maribe Branch, K. Gustafson, K. Nieveen, & P. T. (Eds.), *Design approaches and tools in education and training* (pp. 1–14). Dordrecht, NL: Kluwer.

Elena Nardi
University of East Anglia (United Kingdom)

MARTA MOLINA, REBECCA AMBROSE,
ENCARNACIÓN CASTRO AND ENRIQUE CASTRO

10. BREAKING THE ADDITION ADDICTION

Creating the Conditions for Knowing-to Act in Early Algebra

INTRODUCTION

In the last two decades, within the mathematics education research community there has been a strong interest in analyzing and promoting the integration of algebraic thinking in the elementary curriculum. This curricular proposal, called Early Algebra, aims not just to facilitate the later learning of algebra, but to foster students' conceptual development of deeper and more complex mathematics from very early ages (Blanton & Kaput, 2005; Kaput, 1998). Algebraic ways of thinking are considered to naturally emerge from elementary mathematics and to have the potential to enrich school mathematics activity. This proposal is based on a broad conception of algebra which includes the study of functional relations, the study and generalization of patterns and numeric relations, the study of structures abstracted from computation and relations, the development and manipulation of symbolism, and modelling as a domain of expression and formalization of generalizations (Kaput, 1998).

The Early Algebra view is shared by other researchers (Hewitt, 1998; Mason, Graham, Pimm, & Gowar, 1985) who consider generalization as the root of algebra and highlight the role of algebraic thinking in arithmetic. They argue that arithmetic learning requires students to interiorize generalities about the structure of arithmetic as well as to develop (general) methods to compute and solve problems. All these authors agree that arithmetic teaching needs to provide students opportunities for:
– appreciation of patterns and verbalizing and recording generalizations, as first steps towards symbolically expressing generalizations, and
– becoming aware and making explicit the structure of arithmetic, which is required in order to later be able to use arithmetic structure in algebraic contexts.

In essence, these recommendations argue that algebraic thinking requires children to approach numbers and equations from a structural perspective rather than an operational one, treating expressions as objects instead of processes (Sfard, 1991). These claims are based on the recognition of the poor understanding of relations and mathematical structure that students tend to develop as a result of traditional arithmetic teaching (Kieran, 1989).

In this chapter we use Mason & Spence's (1999) view of mathematical thinking as "knowing-to act" to argue that to engage in algebraic thinking students have to break some habits, and that carefully engineered number sentences along with carefully orchestrated discussions can be a means to that end. Some specific elements that promote students' "knowing-to act" in this context are identified.

RELATIONAL THINKING

We focus our attention on a specific type of algebraic thinking, specifically, *relational thinking*. This term refers to students' recognition and use of relationships between elements in number sentences and expressions (Carpenter, Franke & Levi, 2003; Molina & Ambrose, 2008; Stephens, 2006). When using relational thinking, students consider the sentence and expressions as wholes (instead of as processes to carry out step by step), analyze them, discern some details and recognize some relations, and finally exploit these relations to construct a solution strategy (in a broader context Hejny, Jirotkova & Kratochvilova (2006) named this approach as conceptual meta-strategies). For example, to determine if number equations such as (a) 7+7+9=14+9 or (b) 27+48−48=27 are true or false, instead of doing the computations on both sides and comparing both results, students may solve[1] them by looking at the whole sentence, appreciating its structure (e.g., there are operations on both sides of the equal sign or there are not) and using perceived relations between its elements (e.g., 9 appears in both sides; 7+7 on the left side adds to the other term on the right side, 14; the same number is being added to and then subtracted from 27) as well as knowledge of the structure of arithmetic to determine the truth or falsity of the sentence.

The arithmetic expressions involved have to be considered from a structural perspective rather than simply a procedural one. The expression or object "7+7+9" is compared to the expression or object "14+9" to consider their equivalence rather than acting on each expression to determine its value. This implies a subtle but important change in students' attention from reading the equation from left to right, one piece at a time with a computational perspective, to looking at each side of the equal sign and comparing the two expressions to one another (Mason, Drury, & Bills, 2007).

Mason and Spence (1999) distinguish "knowing-to act" among other less sophisticated ways of knowing: knowing-that, knowing-why, knowing-how, and knowing-about. "Knowing-to act" refers to the use of active knowledge, that is, *"knowledge that enables people to act creatively rather than merely react to stimuli with trained or habituated behaviour"* (p. 136). This type of knowledge, which is contrasted with inert knowledge, is characterized by being transferable to other (new) contexts/situations. This happens because something in the new situation resonates with past experience:

> The state of sensitivity-awareness of the individual, combined with elements of the situation which metonymically trigger or metaphorically resonate with experience, are what produce the sudden knowing-to act in the moment. (p. 147)

In order for this transference to occur, some elements in previous experiences must have been labelled or articulated in some way so that, later, knowledge from that experience can be triggered in new situations.

Mason and Spence use the idea of "knowing-to act" in the moment to discuss the need to get students "unstuck" or to use the knowledge from previous experience in fresh situations where it would be helpful. They note that students too often struggle with mathematics because they fail to apply what they have learned. Here, we are concerned with helping students to "see" number sentences in new ways and to develop flexibility in approaching them by using their previous arithmetic experience. In both cases the aim is to help students develop their awareness of the structure of arithmetic.

One of the hurdles to getting students to use relational thinking is overcoming "habituated behaviour" because they have to resist the impulse to compute. This assumes that the students have some ideas about arithmetic operations that will allow them to employ relational thinking. To use Mason and Spence's (1999) terms, they "know about" addition but do not always act on that knowledge. The presence of the equal sign, or just the presence of numbers and operational signs, leads them to make computations and obtain the numeric values, i.e., what the students think of as the "answer". Students have developed the "know-how" of computing and have been practicing it so much that they fail to notice (or to attend to) other aspects of expressions. To promote relational thinking, the teacher has to help children break their computational habits, in other words, to break the addition addiction, so that they can look at equations/expressions differently. We engaged in a teaching experiment to see if this was possible.

THE TEACHING EXPERIMENT

Design of the experiment

Our teaching experiment shared features of design experimentation.[2] The general aim of this study was to analyze the emergence, development and use of relational thinking in a group of third graders. We worked with a class of 26 eight-year-old Spanish students[3] (12 male and 14 females) from a state school in the region of Granada (Spain). Three of the students received extra support in mathematics at school. The selection of this group of students was due to its availability to participate in the study. We include below a brief description of the design of each session.[4]

Our teaching experiment consisted of six one-hour in-class sessions, over a period of one year. This timeline was chosen intentionally (except from vacation periods) because we wanted (a) our intervention to have a longer effect, (b) to diminish the probability of assessing memory-based learning and (c) to have enough time between sessions to analyze the data of the previous session and take decisions about the next one.

We provided students with number sentences in the context of individual written activities, whole group discussions and individual interviews. We included action sentences (i.e., sentences with all the operations on one side of the equation as in

15+5−3=17) and non-action sentences[5] (i.e, sentences with operational symbols on both sides or with no operational symbols as in 14+6=10+10 and 12=12, respectively). They involved numbers of one, two or three digits and the addition and subtraction operations. These sentences (not all of which were true) were based on the following arithmetic properties and relations:
- commutative property of addition (e.g., 10+4=4+10),
- non-commutability of subtraction (e.g., 15−6=6−15),
- inverse relation of addition and subtraction (e.g., 100+94−94=100; 122+35−35=122),
- compensation relation (e.g., 13+11=12+12; 78−45=77−44),
- unity element (e.g., 0+325=326; 125−0=125),
- inverse element (e.g., 100−100=1),
- composition/decomposition relationships (e.g., 7+7+9=14+9; 78−16=78−10−6),
- relative size comparisons[6] (e.g., 37+22=300; 10−7=10−4; 72=56−14).

Therefore, all the sentences could be solved by using relational thinking as well as by computing. We wondered if certain kinds of sentences were more likely to promote relational thinking than others.

We chose the context of number sentences because (a) it can be a context very rich in patterns (especially patterns related to arithmetic structure), and (b) it is strongly connected to algebraic symbolism. This context offers the possibility to promote the following algebraic elements:
- The conception of expressions as wholes which can be compared, ordered, made equal, transformed, and therefore, the acceptance of lack of closure (i.e. working with expressions without knowing their numeric value or not having it expressed in the sentence).
- The use of horizontal language that traditionally has been more typical of algebra than of arithmetic.
- A two-way interpretation of number sentences as well as their exploration as representations of a static relation between two expressions.

Due to the different objectives of each session (described below) we used open number sentences in session 1 and part of session 2, and true/false sentences in the rest. Open number sentences have proved to be useful for revealing different conceptions and challenging children to reconsider their interpretations of the equal sign, while true/false sentences help to challenge students' computational mindset (Molina & Ambrose, 2008). Students had to complete the open sentences and explain how they solve them. In the true/false sentences, they had to determine if the sentences were true or false and to justify their answers. In the discussions we always encouraged students to share the strategies they used to solve the sentences.

The first two sessions were directed at exploring and extending students' understanding of the equal sign and detecting spontaneous evidence of use of relational thinking. We also analysed students' difficulties in solving the proposed number sentences. From the third session on, we promoted students' *use* and *verbalization* of relational thinking by encouraging working on the same sentence in different ways, asking students for ways of solving the sentences without doing all the computations, and by showing a special appreciation of explanations based on relations. We did not promote the learning of specific relational strategies but the development of a habit of looking for relations, trying to help students to make explicit and apply the knowledge of structural properties that they had from their previous experience with arithmetic. Session 3, 4, 5 and 6 also aimed to identify the strategies used by the students when solving the sentences and to detect and analyze students' difficulties.

We video-recorded the sessions, audio-recorded individual interviews with students and collected the students' worksheets yielding an exhaustive collection of data about the students' thinking while solving the proposed number sentences. In between our in-class interventions, the official teacher faithfully followed a textbook that was mainly centred on computational practice. Some mental computation strategies were introduced, but their use was not practiced more than once. The students never had opportunities to work on non-action number sentences.

Students' strategies

As we expected, at the beginning of the experiment the students demonstrated their computational habit. For example, in the sentence 6+4+18=10+18 a student explained "It is true because 6+4+18=28 and 10+18=28".[7] This student used the vertical standard addition algorithm to compute 6+4+18=28 and 10+18=28. Other students did the computation mentally or by counting. When students computed the numeric value of each side of the sentence, their attention seemed to be focused on the numbers and operations to perform on them, considering each side, or even each operation, separately. They did not provide any evidence of noticing any relation or characteristic of the sentence apart from the numbers in it, the operations that combined them and the presence of the equal sign. On a few occasions students attended to the size of the numbers involved to decide on which computation to perform first. In these few cases they used relations to inform how to address the computation. Following this computational habit, when asked to provide a different explanation for the same sentence, students proposed a different order in which the expressions could be computed.

However, relational thinking became relatively frequent as we started to promote the use of this type of thinking. In sessions 1 and 2 we detected the first two examples that were displayed:

In $12 + 7 = 7 + \square$, a student explained that the missing number was 12 because "It is the same number [R: Is it the same number? What did happen?] They just changed the order of the numbers".

Another student wrote the following explanation in the sentence $9 - 4 = \square - 3$: "I have used my mind to do it because ...

$$9-4=8-3$$

... and it gives the same". Her writing under the sentence suggests that she appreciated that if 1 is subtracted from 4 you get 3 and, therefore, she subtracted one from 9 to get the answer, 8.

These explanations demonstrate "knowing-to act" because the students used their previous arithmetic knowledge in a fresh situation. They broke their computational trained behaviour and used their natural powers to discern patterns and recognize relations. Both strategies were specifically constructed by the students attending to the particular characteristics of these sentences.

These students "knew that" the equal sign is used to express equivalence between numeric expressions, i.e., sameness of numeric value. They "knew how" to compute addition and subtraction expressions (at least involving numbers of less than 5 digits). Therefore, they could successfully address this task following a computational approach but also could use a creative/non-trained approach as in the examples shown above which make the most of the special design of the sentences (rich in relations).

When students used relational thinking, their attention was directed to particularities of the sentences – presence of zero, the particular operational signs involved, the presence of operations on both sides – and relations between their terms such as sameness, lack of sameness, difference of one unit between terms and big differences of size between numbers. These observations resonated with their previous experience about, for example, "adding or subtracting zero", "adding and subtracting the same number", "changing the order of the addends in a sum", "the effect of adding/subtracting on the size of numbers". In this way, some previous (implicit or explicit) "knowledge-about" arithmetic structure was flexibly applied in this new context (see further examples in table 1).

Occasions for relational thinking

We distinguish three occasions where we observed relational thinking: without computing, while computing and after computing. In the first case, students approached the sentences by attending to its structure and detected particular characteristics or relations between its elements which they used to conclude their answer. They did not need to perform any computation (e.g., in 75+23=23+75, a student explained *"True because there are the same numbers in one operation and in the other"*; she didn't do any computation).

Table 1. Examples of students' explanations evidencing use of relational thinking

Sentences	Examples: Students' Explanations
$122 + 35 - 35 = 122$	"True because if we add 122 to 35 and we take it away, it is as if we don't add anything."
$7 + 7 + 9 = 14 + 9$	"True. I did it by adding seven and seven…. which is fourteen. The same than there [right side]. Nine, the same than there [right side] too."
$13 + 11 = 12 + 12$	"True because you subtract one to the twelve and you give it to the other twelve, and you get what it is there [i.e., the expression on the left side]."
$75 - 14 = 340$	"False because 75 minus 14 is less, it cannot be a bigger number."
$11 - 6 = 10 - 5$	"True because if eleven is higher than ten and you subtract one more than five, you get the same."

In other cases, students initiated some computation to obtain the numeric values of both sides but, suddenly, abandoned the computation and changed their approach after appreciating some characteristic of the sentence or some relations between its terms, not previously noticed. Initiating the computation served to make the student aware of the composition of the sentence and pay attention to each of its elements. For example, in the sentence 51+51=50+52 a student provided the following explanation: *"It is true, because fifty-one plus fifty-one is one hundred and two, but fifty-one, if you subtract [one], fifty, you can add [it] to the other fifty-one, one more, and you get fifty-two"*. He initially computed the expression on the left side, but then appreciated a compensation relation between the operations in both sides and used it to conclude his answer, without computing the expression on the right side. In the sentence 75+23=23+75, a student began by writing the numbers in a vertical format and then did not even start computing as she explained, *"It is true because it is the same"*. Her explanation suggests that she appreciated some sameness between the expressions on each side of the equal sign which allowed her to conclude that the sentence is true without knowing the numeric values of each side.

In other cases, students first solved the sentence by computing and comparing the numeric values of both sides, and afterwards explained another way of concluding the truth or falseness of the sentence which was based on some noticed relations (e.g., in the sentence 7+15=8+15, having computed the addition on both sides, a student explained "False because you don't get the same [result] and, also [because] seven is lower").

These examples illustrate how shifts in attention happen instantaneously. Mason & Spence (1999) related this to a "bolt of lightening" when patterns emerge suddenly and all of a sudden students "know-to act". Relational thinking does not always precede or is opposite to computation. As the above examples show, it can occur during computation as students have an insight about their computation, allowing them to abandon it. Working on the computations was helpful for some students to become aware of the structure and components of the sentence and perceive relations between them that they may use to solve the sentence. Their "knowing-to act" demonstrates that during the computation process they had some free attention to attend to these details that resonated with past experience.

Most students demonstrated some use of relational thinking, in each of these three ways, at some point during the teaching experiment; however, it was alternated with a computational approach. They advanced in this regard throughout the sessions being strongly influenced by social appreciation of explanations based on this type of thinking.

Students' "knowing-to act" during the sessions

Of the 26 students, six students used relational thinking frequently from the third session on, and three of them did so in all the types of sentences considered (according to the arithmetic relation used in their design). Another ten students evidenced some use of relational thinking occasionally. It was based only on some specific relations but all of them appreciated sameness or lack of it (the most basic relations). Ruben was one of the ten students whose "knowing-to act" was based on specific relations: sameness and composition. He noticed repetition of numbers in both sides of the sentences 75+23=23+75 and 18−7=7−18 that he used to conclude their truth, and in the sentence 6+4+18=10+18 (in session 6) he appreciated a composition relation that he used: "True because 6+4=10+18= 10+18".[8] In his written explanation he incorrectly used the equal sign to chain a sequence of operations. We suspect that he was able to apply relational thinking to this sentence because he noticed that the 18 was the same on both sides leading him to look for a relation between 6 + 4 and 10.

Another example is the case of Maite. Her approach was computational in all the sentences except from her work in 18 − 7 = 7 − 18, 75 + 23 = 23 + 75 and 7+15=8+15: "True because both are the same", "True because it is equal" and "False because it is almost equal but it is not equal", respectively. In all these sentences she previously did the computation of the right side or wrote it vertically and stopped before computing. She displayed a tendency to calculating when approaching the sentences but in these three non-action sentences her attention was not completely taken by the computations, allowing her to recognize some sameness between both sides of the sentences.

The other six students showed limited use of relational thinking. Three of them noticed sameness between the terms in a sentence. In the other three it was based on noticing an instantiation of the property $a-a=0$ or the properties of zero as identity element. They also applied the restriction of subtraction in the natural numbers to operations where the minuend was not lower than the subtrahend (i.e., $a-b$ when $a < b$).

Only two or three students never provided evidence of using relational thinking. One of these students was Beatriz. As shown in the examples provided in Table 2, she typically computed the numeric values of the expressions on each side of the equal sign by using the vertical standard addition and subtraction algorithms. Like Beatriz, the other two students who did approach all the sentences computationally displayed some difficulties in computing and sometimes did not respect the structure of the sentences and performed computations which combined numbers from different sides of the equal sign (e.g., see Beatriz's work in the sentence 6+4+18=10+18). We conjecture that their lower mastery of computational methods did not allow them to have free attention while computing to perceive relations between the terms and probably also to follow and benefit from their peers' explanations in the whole group discussions. It also caused them to struggle to make sense of non-action sentences that were not familiar to them.

Table 2. Beatriz's responses to some true/false number sentences

Number sentence	Responses*
18 − 7 = 7 − 18	False because it doesn't give the same result. (She computes 18 − 7 = 11 and 7 − 18 = 19 by using the vertical standard subtraction algorithm.)
75 − 14 = 340	False because I added and it doesn't give 340. (She computes 75 − 14 = 81 by using the vertical standard subtraction algorithm.)
17 − 12 = 16 − 11	True because I subtracted 17 and 12 and later I subtracted 16 − 11.
122 + 35 − 35 = 122	True because I added and then I subtracted the result [that] I got. (She computes 122 + 35 = 157 and 157 − 35 = 122 by using the vertical standard algorithms.)
6 + 4 + 18 = 10 + 18	False because I added 6 + 4 and I added the result to 10 + 18 (She computes 4 + 6 = 10, 6 + 4 + 18 = 118, 118 + 10 + 18 = 146 by using the vertical standard addition algorithm)

* Within parentheses we describe the students' computations done in the worksheet.

Role of number sentences

In addition to creating an atmosphere where "knowing-to act" was valued and having discussions about children's ways of looking at those sentences, the type of true/false number sentences considered were an important element in helping students "knowing-to act in the moment."

Number sentences that include zero relations ($a + 0 = a$; $a - 0 = a$; $a - a = 0$) seemed to be effective tasks to interrupt students' habituated behaviour. In these sentences the use of relational thinking was facilitated by not having to relate both sides of the sentences. Even those students who were the most likely to compute (although not always) tended not to do so on these types of sentences. Only one student wrote the operation 125 − 125 vertically to determine if the sentence 125 − 125 = 13 was true or false. In this case the size of the number seemed to be a problem for him to conceive 125 as a number.

Sentences involving the commutative property also seemed to interrupt habituated behaviour for students who otherwise computed, although not always. In each session, more "knowing-to act" than computational approaches were displayed in this type of sentence. In the discussion of session 3, none of the students solved the sentence 10 + 4 = 4 + 10 by computing. Various students claimed loudly that it was true and explained: "they had turned around the numbers". In sessions 4 and 6, only 5 and 8 students, respectively, solved the sentence 75 + 23 = 23 + 75 by computing the numeric value of each side, while 15 out of 22 students in session 4 and 9 out of 20 students in session 6 determined the truth of the sentence by "knowing-to act".[9] They explained that the numbers were the same and some mentioned the change of order.

In some cases (7 out of 24 students, in both sessions 4 and 6), the appreciation of sameness in sentences such as 18 − 7 = 7 − 18 lead students to reason erroneously as result of having over-generalized the commutative property of addition to the case of subtraction (e.g., "True because eighteen minus 7 and the other is the same, and if it is the same they are equal"). Students' "knowing-about" this type of expression may have been limited by their lack of experience. They had been told that subtraction cannot be performed when the minuend is lower than the subtrahend. Therefore, encountering an expression in which they couldn't compute may have led some of them to assume that the commutative property could also be applied here.

In the sentences based on composition/decomposition relation (e.g., 6 + 4 + 18 = 10 + 18) as well as on the inverse relation of addition and subtraction (e.g., 122 + 35 − 35 = 122) half of the students proceeded computationally while the other half used relational thinking. However, in the latter we detected more use of computational approaches when the sentences included small numbers.

In the sentences based on "relative size comparisons" initially, during the whole group discussion of session 3, students evidenced both approaches but computational approaches became more frequent in sessions 4 and 6. This tendency was specially appreciated in the action sentences considered (72 = 56 − 14; 75 − 14 = 340; 37 + 22 = 300) probably because they did not include equal numbers on both sides while the others (10 − 7 = 10 − 4; 7 + 15 = 8 + 15) did.

The sentences based on the compensation relation (e.g., 53 + 41 = 54 + 40) were the ones least frequently approached relationally, especially those involving subtraction (e.g., 9 − 4 = 8 − 3). In the discussion of the sentences 51 + 51 = 50 + 52 and 13 + 11 = 12 + 12 in session 3, three students' "knowing-to act" became evident. However, despite having shared these explanations with the whole group, in the later sessions only four students displayed relational thinking in sentences based on the compensation relationship.

Clearly students used relational thinking more readily in some types of sentences and use of relational thinking was sporadic for most students. Except for three to six students, it was not the case that they had an insight which led them to be on the look-out for relations. Instead, sometimes the relations in the sentence jumped out at them and other times they did not. The data above indicate that when at least some of the numbers on each side of the equal sign are identical, students are inclined to notice relations.

TO CONNECT WITH THE READER'S EXPERIENCE

As John Mason usually does, now we ask the readers to try an example themselves so that they experience the ideas we are trying to describe. We invite you to solve this algebra example $x^2 - 2x = 4x - 8$ before going on reading.

In this equation you might have factored both sides or you might have subtracted $4x$ from the left side, and added 8 to both sides. What one decides to do depends on what one notices. If you noticed that it was quadratic, you might realize that there will be two solutions to this equation. If you noticed that $(x - 2)$ is a factor of both expressions, you might have guessed that 2 is one of the solutions. This factorization would avoid having to apply the formula for solving quadratic equations that constitute the trained behaviour in this context. You might have addressed the equation looking at its structure and its components and searching for relations between both sides and its terms that inform your approach. You might have initiated some manipulation before noticing any particular characteristic or have noticed them when checking your solutions after using the formula for quadratic equations.

The flexible thinking involved in "knowing-to act in the moment" would allow you to weigh your options and gain insight into the nature of the equation. Attending to the structure of the equation allows you to enrich your knowledge of the equation at hand in addition to informing your selection of a solution strategy. In addition possible discussions starting from this "knowing-to" may lead to an interesting inquiry about other equations that are similar in some way (e.g., what can we change in the sentence so that 2 is still a solution and the other one is 4? or –7? or $^9/_5$?).

Teaching students to follow a series of steps, as in teaching solving equations, does not help them to learn algebra, or any area of mathematics, properly because there are always exceptions to any given series of steps they might try to memorize, and more important, this learning does not allow them to establish connections with other mathematical concepts, therefore limiting their understanding.

CONCLUSIONS

There are some mathematical situations when students get stuck when they do not "know-to act." The tasks we proposed did not have this limitation because students could calculate to successfully obtain an answer. This fact made the tasks accessible to all students but caused the necessity of breaking students' trained behaviour and changing their disposition when approaching the sentences.

Asking students to determine and justify the truth or falsity of number sentences in a classroom atmosphere where the focus was not on numeric results nor on calculations but on recognising and expressing relationships was successful in promoting the display of "knowing-to act" and altering how students attended to number sentences. Some students didn't "know-to act" until they encountered relational strategies when listening to their peers; in Mason and Spence's (1999) words until they were aware of a possibility to act. Computational approaches were the most familiar strategies for them (if not the only for some), so the sharing of the students' strategies and the teacher's special appreciation of relational strategies was essential to support "knowing-to act".

We were completely successful in developing a habit of looking for relations in the case of three to six students. Most of the students became aware of the existence of non-computational strategies and when some particularities of the sentence resonated with their previous experience, usually some sameness, it allowed their attention to focus on the noticed relations and triggered their knowledge-about them to solve the sentence.

Some students never used relational thinking probably because they needed to devote all of their attention to the calculation at hand, so there was insufficient free attention for any metonymic trigger or metaphoric resonance between the elements in the sentences. Some may have needed some assistance or teaching to help them to investigate the effect of some arithmetic relations such as composition/decomposition and compensation in the numeric value of expressions. One particular student required further experience with big numbers to be able to conceive them as numbers.

Although the results show that six sessions were not enough to fully reach our objective, they prove that true/false number sentences of the forms described here can be fruitful in helping students break their "addition addiction". These activities provoked a transformation in the structure of students' attention and made it more aligned with the requirement of algebraic thinking: attention to the structure of the sentence and to relations between its terms. Students sometimes see relations after starting to compute, supporting Mason and Spence's assertion that shifts in attention can happen quite rapidly. Students tend to notice some relations (zero and sameness relations) more readily than others so teachers need to be prepared with a variety of sentences to support students at different levels of development to use relational thinking. In some cases the presence of big numbers eased the appreciation of relations.

Students need to appreciate that "knowing-to act" (in this context, relational thinking) requires shifts in attention, a playful approach to a problem, and some creativity. They have to avoid reacting to stimuli with habituated behaviour and reserve habituated behaviour until after they have considered alternative possibilities. This disposition is important to all areas of mathematics but we have observed that discussing true/false number sentences provides some unique opportunities for students to recognize the value of shifting attention.

NOTES

[1] In this paper we use the expression "to solve" the true/false sentence to briefly refer to determine if the sentence is true or false.
[2] For further information see Molina, Castro, & Castro (2007).
[3] The results will only refer to twenty-five students, as the other student only attended session 1 and 4, and he did not solve the written assessment of session 4.
[4] For further information about the justification of the design of each session see the first author's PhD thesis at http://cumbia.ath.cx:591/pna/Archivos/MolinaM07-2822.pdf.
[5] This classification of number sentences comes from Behr, Erlwanger, & Nichols (1980).
[6] Here we consider sentences in which students can determine the validity of the sentence by attending to the size of the numbers involved and using knowledge of the effect of operations on the size of numbers.
[7] All the examples of students' explanations provided in this paper have been translated from Spanish to English.
[8] Some children over-generalized the commutative property; an issue addressed in the next section.
[9] In sessions 4 and 6 there were two and three students, respectively, whose approach could not be identified due to lack of details in their explanations and written work.

ACKNOWLEDGMENT

This study has been developed within a Spanish national project of Research, Development and Innovation, identified by the code SEJ2006-09056, financed by the Spanish Ministry of Sciences and Technology and FEDER funds.

REFERENCES

Behr, M., Erlwanger, S., & Nichols, E. (1980). How children view the equal sign. *Mathematics Teaching, 92*, 13–15.
Blanton, M. L., & Kaput, J. (2005). Characterizing a classroom practice that promotes algebraic reasoning. *Journal for Research in Mathematics Education, 36*(5), 412–446.
Carpenter, T. P., Franke, M. L., & Levi, L. (2003). *Thinking mathematically: integrating arithmetic and algebra in elementary school*. Portsmouth, NH: Heinemann.
Hejny, M., Jirotkova, D., & Kratochvilova, J. (2006). Early conceptual thinking. In J. Novotná, H. Moraová, M. Krátká, & N. Stehlíková (Eds.), *Proceedings of the 30th conference of the international group for the psychology of mathematics education* (Vol. 3, pp. 289–296). Prague, CZ: Program Committee.
Hewitt, D. (1998). Approaching arithmetic algebraically. *Mathematics Teaching, 163*, 19–29.
Kaput, J. (1998). *Teaching and learning a new algebra with understanding*. Dartmouth, MA: National Center for Improving Student Learning and Achievement in Mathematics and Science.
Kieran, C. (1989). The early learning of algebra: A structural perspective. In S. Wagner & C. Kieran (Eds.), *Research issues in the learning and teaching of algebra* (Vol. 4, pp. 33–59). Reston, VA: Erlbaum.
Mason, J., & Spence, M. (1999). Beyond mere knowledge of mathematics: the importance of knowing-to act in the moment. *Educational Studies in Mathematics, 38*(1–3), 135–161.
Mason, J., Drury, H., & Bills, L. (2007). Studies in the zone of proximal awareness. *Plenary presented at the 30th Conference of the Mathematics Education Research Group of Australasia*. Hobart, TAS.
Mason, J., Graham, A., Pimm, D., & Gowar, N. (1985). *Routes to, roots of algebra*. Milton Keynes, UK: The Open University Press.
Molina, M., & Ambrose, R. (2008). From an operational to a relational conception of the equal sign. Thirds graders' developing algebraic thinking. *Focus on Learning Problems in Mathematics, 30*(1), 61–80.

Molina, M., Castro, E., & Castro, E. (2007). Teaching experiments within design research. *The International Journal of Interdisciplinary Social Sciences, 2*(4), 435–440.

Sfard, A. (1991). On the dual nature of mathematical conceptions: Reflections and objects as different sides of the same coin. *Educational Studies in Mathematics, 22*(1), 1–36.

Stephens, M. (2006). Describing and exploring the power of relational thinking. In P. Grootenboer, R. Zevenbergen, & M. Chinnappan (Eds.), *Proceedings of the 29th annual conference of the mathematics education research group of Australasia* (pp. 479–486). Canberra, ACT: MERGA.

Marta Molina
University of Granada (Spain)

Rebecca Ambrose
University of California, Davis (United States)

Encarnación Castro
University of Granada (Spain)

Enrique Castro
University of Granada (Spain)

ALF COLES

11. TOWARDS AN AESTHETICS OF EDUCATION

"Yet one *is* that with which one concerns oneself". (Heidegger, 1962, p. 368)

SCIENCE AND THE AESTHETIC

I am a teacher and a researcher. In my research work I find myself caught between the dilemma of wanting to 'find something out' that might be useful to others, and my intuitions from teaching that, the only person I can hope to try and change is myself. I associate the former view of education with a 'scientific' stance – the hope that researchers may be able to build on each other's work and that, as teachers, we may be able to 'stand on the shoulders of giants' in our work in classrooms. The latter view I associate more with notions of teaching as an 'art' – that learning to teach is an aesthetic endeavour (which I take to mean, one involving personal transformation) to become more and more sensitive to what happens in a classroom. This chapter can be read as an attempt to reconcile these two views within my own work.

I first came across the notion of a 'Science of Education' through joining, in my first year of teaching, the (now disbanded) Working Group of the same name funded by the UK's Association of Teachers of Mathematics (ATM). The name of the group came from Gattegno's 1987 work, in some respects the culmination of his life work. Gattegno urged all teachers to become scientists of education through the application of 'watchfulness'. He felt that all sciences start with an awareness (e.g., Chemistry began with the awareness that materials could be analysed according to their molecular composition). The foundation of this new science of education was to be the awareness of awareness itself; in the words of Gattegno's aphorism: "Only awareness is educable".

In similar vein, Mason's (2002) 'discipline of noticing' starts from the premise that practitioners in any domain want "to be awake to possibilities, to be sensitive to the situation and to respond appropriately" (p. 7). I read this as confirming Gattegno's notion of the importance of awareness, and have always seen John Mason as an archetypal Scientist of Education.

The ATM Science of Education Group had certain well established disciplines and ways of working, although when I first joined it was not at all clear to me what these were. Despite the name, there was seemingly no set of 'scientific' procedures I could simply follow in order to become a better teacher. I wanted my students to be autonomous, passionate and successful, yet in beginning teaching the reality of my classroom was far from these ideals. To engage with the conception of a Science of Education embodied by the members of the ATM Group has, in my experience, entailed personal transformation (see Brown & Coles, 2008, for a description of this journey).

In Gattegno's conception of science, and in Mason's writing, knowledge is seen as lived, and cannot be disassociated from the knower. Dewey (1929) in similar vein called for a Science of Education which would not be seen in opposition to the view that teaching is an Art.

> It is very easy for science to be regarded as a guarantee that goes with the sale of goods rather than as a light to the eyes and a lamp to the feet. (p. 15)

> If we retain the word "rule" at all, we must say that scientific results furnish a rule for the conduct of observations and inquiries, not a rule for overt action. They function not directly with respect to practice and its results, but indirectly, through the medium of an altered mental attitude. (p. 30)

The notion of 'an altered mental attitude' I take as being synonymous, in Gattegno's language, with the gaining of new awarenesses, or for Mason 'noticing' the new. I read these quotations as implying that even the 'accepted' findings of 'hard science' have to be re-created in the practice of each new scientist. We perhaps can be blinded to this personal dimension of science through the embodiment and embedding of ideas within a culture, meaning that practices can become accepted unthinkingly, in the form of habits.

Mason (2002), unsurprisingly, takes such a personal view of the progress of education research. His 'discipline of noticing' explicitly foregrounds personal transformation as the key aspect of taking part in research. He coined the phrase 'researching from the inside' (Mason, 2002, p. 204), consistent with Gattegno's sense that the science of education is about the development of awareness.

> What is important in qualitative research in general, and in researching from the inside in particular, is not the validity or accuracy of the description, but the effect, the action that the description sets up inside others. (Mason, 2002, p. 229)

The importance of something being defined by its effect inside others is a feature more usually associated with an artistic (or aesthetic) rather than scientific experience. Again, the personal and transformative aspects of the process of research are to the fore. However, to use the word 'aesthetic' in the context of education research would entail an expansion of meaning to areas outside the Arts. I offer two descriptions below, either of which I would take to be an aesthetic experience:
– an experience in which I am engaged in a task to the point where I lose self-consciousness; all my attention is absorbed in what I am doing;
– an experience of transformation – e.g., I gain an awareness, my perception is altered.

Such an extension of what constitutes the aesthetic is not a new idea. Dewey (1934) stated the aim of his book was:

to restore continuity between the refined and intensified forms of experience that are works of art and the everyday events, doings, and sufferings that are universally recognized to constitute experience. (p. 2)

He wanted to broaden the range of what might be considered 'Art' and hence the range of activities in which we might say we are aesthetically engaged, to include:

[t]he intelligent mechanic engaged in his job, interested in doing well and finding satisfaction in his handiwork, caring for his materials and tools with genuine affection (p. 4)

I take it that teaching and research can be done with similar care and affection.

Aesthetic and scientific viewpoints offer two contrasting lightings on the complex whole that is education. The aesthetic dimension privileges the personal and transformative; the scientific privileges the social and verifiable. I cannot help feeling that an 'Aesthetics of Education' is a more telling description of the line of enquiry traced through Dewey, Gattegno and Mason, than a 'Science of Education', although of course both aspects are important.

AESTHETICS AND RESEARCH METHODOLOGY

I have frequently had the experience of reading a research report or hearing a speaker, who has meticulously set out theoretical frames, methodology and methods, and yet still being left in the dark as to what they actually did to arrive at their abstractions. It seems easy to sign ourselves up to a theoretical position and gloss over issues of whether the way we act is consistent with such beliefs. An implication of taking an aesthetic view of education for research would be a refusal to duck issues of personal change undergone by the researchers and relevant aspects of the researchers' histories. I believe, as a researcher, I need to give some account of where I place my attention when, for example, analysing data. How we use ourselves matters. It makes a difference to what I see from the back of a classroom if, say, my intention is to record as much detail as possible of what the teacher says, or if I have a list of characteristics I am looking out for.

What I am articulating here is, in part, my belief in the importance of what Bruner (1990) termed a 'culturally sensitive psychology':

(which) is and must be based not only upon what people actually do but what they say they do and what they say caused them to do what they did. (p. 16)

Reading the quotation in reflexive spirit, I take this as a comment about the central plank of the research process, i.e., the need to account for how our beliefs inform and have informed what we do. However Bruner goes on to state the need to reverse this direction of accounting, and muses:

... how curious that there are so few studies that (ask): how does what one does reveal what one thinks and believes? (p. 17)

An aesthetic take on the reporting of qualitative research would have to address the question above by giving some account of *how* we do what we do, in the doing of research.

Into the classroom

The tension between the aesthetic and scientific resurfaces again, for me, when it comes to studying classroom interaction. As a teacher, I have long been convinced of the importance of classroom discussion; in terms of generating engagement and commitment to activities; in terms of shaping classroom culture; and in terms of driving individual learning. In my Masters' dissertation, I analysed the listening of two teachers in the UK, during spells of whole class discussion, and developed from Davis (1996) the notion of 'transformative listening' (Coles, 2001). This notion aims to say something about the state of mind of the teacher (and student) when statements are not evaluated or even interpreted, but instead given space so that they may be heard by others, may alter the direction of a lesson, may alter the listener's ideas. When I am able to be 'mindful' or 'alive to the present' moment (Thera, 1996, p. 11) with a class and listen in such a transformative manner, it feels as though my and the students' actions are in some kind of 'flow' (in the sense described in Csikszenhmihalyi, 1997).

In contrast, evaluative listening is signalled by a judgment (right/wrong) made on what someone else has said. The judgemental or evaluative stance signals a separation and refusal to engage in another's perspective.

In between these two forms of listening, Davis (1996) introduced the notion of 'interpretive listening'. In this mode, I am interested in making sense of what someone else has said, and recognise that they may be interpreting events from a different viewpoint; yet I am not open to transforming my own perspective.

The aesthetic dimension is present in these interests, in that I am concerned with times of engagement and personal transformation. However, in analysing classroom discussion, I am now wary of bringing well defined categories in advance of a detailed study of the data. While a concern with listening and how we respond to each other in discussion is part of who I am and therefore what I notice, I attempt to approach my data more gently, staying open to more delicacies than might be apparent in any three part categorisation.

I believe, with Gattegno and Mason, that the research tool for education is awareness, and want to say something about how I use this tool and where I place my attention. In working with video, audio and transcript data I will spend time dwelling in the detail of what is said, listening to the same small section again and again. During this process, I will note down any connections as they occur to me, but in general I pay close attention to the details of the words being said, trying to avoid too quick a move to categorisation. There are some key questions I keep in mind while considering each turn in a discussion: what pattern does this turn follow; what pattern does this turn break? I am not looking for causal connections, but I am interested in what breaks in patterns occur following different interventions by students and teacher. I share my emerging interpretations with the teachers involved, either individually or as a group in order to triangulate my analysis and will return to study the data again in the light of their comments.

I present below excerpts of two sections of transcript, one from a lesson by a teacher in the school in which I work, and one from a meeting of Mathematics teachers in the same school (I was the Head of the Mathematics Department at the time and was chairing the meeting) analysed, as described above, through looking for pattern. In both transcripts I argue that there is evidence of individual transformation and I conclude by drawing parallels between the two dialogues and suggesting possible implications.

Classroom discussion

The discussion below took place in a mathematics lesson of a mixed ability class of 11–12 year olds in the UK. It is a very early lesson in the year of this, their first year at secondary school. The school has an intake well below national averages in terms of students' prior attainment in mathematics. The class had been working on a problem (called '1089') in which an issue of disagreement had arisen. The problem involves students in a multi-stage process and a difference in end result occurs depending on whether you keep a zero to the left of a number or not when part way through the process.

Four responses to students by the teacher are below; these are typical examples of what the teacher might say in whole class discussion, both in this and other lessons:

> T/S indicates whether a teacher or student is speaking. Square brackets are used to indicate gestures or notes from transcription. Three dots indicate a pause of more than 2 seconds. The dialogue has generally been put into sentence form with standard punctuation as an attempt to give a limited indication of intonation. A slash (/) indicates interrupted speech.

29:45	T	So, we've got a couple of suggestions, yes go on
31:12	T	Go on, anybody got any comments? Natasha
31.39	T	Go on, can you say what your conjecture is
32.55	T	Okay, any thoughts on that? Johnny

In her responses, the teacher consistently comments to students *about* what they have said, rather than responding directly *to* what is said. This practice constitutes a striking break in pattern compared to other classrooms I have recorded in that there is no evaluation by the teacher about the mathematical content of what is said (signalled e.g., by a 'yes' or 'no' or a word of praise/blame).

At 30.10 a student initiates a new direction in the discussion by asking permission to offer a suggestion, implicitly negotiating the terms on which dialogue takes place in this classroom. In fact, I learnt in jointly viewing this video with the teacher, the student had been 'primed' during an earlier part of the lesson; before this discussion the teacher had said to her individually that what she had done might be something good to share with the rest of the class.

| 30:10 | S4 | Could I say what I was going to say about number 2? |

Following this comment the content of the classroom dialogue is directed by students (see below), the teacher hardly comments, except on occasion to invite another student to speak. Students refer to what others have said. The next five student contributions follow. The students are consistently operating at the level of generalization with the teacher in the role of chair or facilitator again making comments *about* the dialogue, rather than within the mathematics.

31:02	S5	You don't necessarily have to put the zero in front of the 99, so if you changed it round [unclear]
31.20	S6	Um, um, on that one if no one never left off the zero all the answers would be, all the answers would be, um, um one thousand and eighty nine, so um that would [unclear] our conjectures
31.41	S6	So, it doesn't really matter what number it is so long as, so long as 'a' is bigger than 'c' um, you get 1089
32.44	S7	When you actually do the sum first and you come up with 099, I reckon you should actually leave the 'oh' on because that's the answer you come up with. It's like doing 4 take away 2 and saying Oh, I'm not going to use 2, I'm going to use 3. It's changing the sum.
33.01	S8	No, because even if you done something to the zero it's still the same number, ninety nine, and in the sum you're only using ninety nine and not nine hundred and ninety as well as 099 (TA gestures to another student)

Students are responding to each other and pooling ideas. The evidence from the transcript is that, in this very early lesson of the year, students have picked up the message already that what is valued in this classroom is what they have noticed within the mathematics; nine different students speak in the 4 minutes of this dialogue – a third of the group.

The five student comments quoted above are all at the level of generalization. Students have noticed a pattern which they report. Following these generalizations, the teacher makes the following comment:

| 33:20 | T | So, I think what you need to be clear about as somebody working on mathematics as a mathematician; you need to decide what you're going to do and be consistent about that. Amy, you're feeling strongly you need to leave the zero in; you might not feel strongly you need to keep the zero in. Natasha's conjecture might not apply to your rules. So when we're making conjectures we need to be clear about the rules we're using, I suppose; so you might want to add on to your conjecture; Natasha's, this is when you leave the zero's or something like that. Could anybody adapt this conjecture if they didn't want to leave the zero in? |

The teacher here is able to make a comment about how mathematicians operate 'we need to be clear about the rules we're using' and invites students to be explicit about what rules they will use. Laurinda Brown and I have, in previous research,

TOWARDS AN AESTHETICS OF EDUCATION

labelled such utterances *about* students' mathematical work 'metacomments' (e.g. see Brown & Coles, 1999). Comments at this level, it is our conjecture, are significant phenomena in sustaining the mathematical orientation of a classroom community and giving students a purpose to their work.

Teacher discussion

A practice I have established in the department where I work is the joint viewing of small excerpts of video recordings of each of us teaching. I set up a camera at the back of the classroom, focused on the board at the front, and leave it running for the lesson. I will then select a small section of classroom dialogue to work on as a group. I always look for a section where there is a whole class discussion taking place and one where there are some significant contributions from students. By significant I mean that, for example, a disagreement between two of more students has the space to be aired; or that there are sections of dialogue that do not follow a T – S – T – S pattern; or that a student contribution seems to alter the direction of the lesson.

We have an established discipline, as a department, when working on sections of video. Having shown everyone the small clip (usually not more than three minutes long) we will begin by trying to reconstruct what happened. This disciplined way of using video recordings of lessons with groups of teachers was described as part of an Open University video research project (Mason, 1985). In keeping with this discipline, I am very clear during the reconstruction phase, that I will interrupt anyone who begins on an interpretation of events, and focus the group back on the task of agreeing the detail of what we can observe happened.

The distinction between interpretation and observation echoes the important notions of 'account of' and 'account for' (Mason, 1996, pp. 23–24). As Mason suggested in coining these phrases we begin with the 'account of' before entertaining any 'account for'. The dialogue below begins at the point the group of teachers move away from reconstruction into an analysis of what happened. The time indicates that twenty minutes have been spent in watching and then reconstructing the three minute clip of video.

The more I work with video in this way, the more I become convinced of the need for a period holding off from interpretations (or accounts 'for') events. It is as though without first entering the detail of what happened I am not able to get beyond my initial snap judgements of events. Without losing myself first in sorting out what actually was said or done, I am unable to see anything different, new or surprising in a section of video.

In the teacher meeting I now report on, we had been working on the exact section of video from which the classroom transcript above was taken.

Analysis

Compared to the classroom discussion, this teacher dialogue is much more 'lightly' chaired (by TC, who is me); TC does not respond to every utterance. Although the practice of viewing videos of lessons in well established in the department, there are three teachers in this group who are new to this way or working.

A pattern consistent with the classroom data is that there are no negative evaluations of each other's comments in this dialogue, and the teachers are mostly operating at the level of generalisation. For example, the first comment by a teacher, after TC's invitation to shift the discussion to what teaching strategies had been employed (i.e. 'accounting for'), was a comment about the whole section of classroom dialogue:

| 20.38 | TE | She didn't personally get involved in what they were doing ... because I know I would have sort of 'look at this', you know, I would have commented [unclear] I just know I would have, especially when it got down to the bit where it was 99 or 099. I just know I would have jumped in. |

There is evidence, for me, of TE gaining an awareness: 'I just know I would have jumped in'. This whole utterance was said in an unusually slow and low voice. A subsequent interview with TE confirmed my sense at the time that this was a significant realisation and one that, TE reports, has changed her classroom practice, i.e., she reports now working on being much more neutral in responding to student comments than she was before.

Immediately following this comment from TE, a second teacher (TB) offers what I read as a personal reflection on the issue of whether as a teacher I 'jump in' or not:

| 21.02 | TB | It feels like a battle doesn't it between wanting to say: 'yes, so what you're saying is de de de' ... um, and instead of saying 'that's a really good thing', saying 'yes, you're completely correct'; she's kind of, you're not saying anything; your offer that you're writing is on the board; you're kind of being there more as a support |

A few minutes later, TB again follows another teacher's comment with a personal reflection:

| 23.28 | TB | Because I think I remember focusing on that after a lesson that I had with this and I remembered, I think I remembered hearing you say 'and now we need to make a decision, a mathematician needs to make a decision'; and I did use that and I said, 'a mathematician needs to make a decision' and I, but I think that I decided what to choose |

A third teacher (TD) follows a train of thought through three contributions over the space of 4 minutes of discussion, and at 24.40 makes a connection, an awareness linked to the teacher in some sense sorting out the students' difficulties and in some sense not.

TOWARDS AN AESTHETICS OF EDUCATION

| 24.40 | TD | Because there was this issue that belonged to the class, and I think you said 'we need to sort this out', um, and if it was me I would have chosen one of the options for them, um, and that's sorting it out for them. I think in some ways you sorted it out for them as well, but gave them the option of putting conditions, so you didn't make the choice but you did sort out the issue ... I don't think ... they didn't sort out the issue, they were wondering what can we do with this, we need to sort this out as well; and I don't think it was sorted out by the pupils, I think you sorted it out, in putting conditions |

It is hard to read the teachers in this discussion as picking up on what each other has said. While, as stated above, there is evidence of personal transformation, this seems to come from the engagement with the video rather than building on each other's comments. I read the dialogue as following 2 or 3 related but parallel strands. There is TE's comment in 20.38, which, on its own, seems to have signalled a beginning of a shift in practice. TE hardly speaks again. Then there are the offers of TB, who starts off with the intriguing phrase 'It feels like a battle' in 20.12 and seems to follow a strand of her own reflections on her practice. Finally there are the comments of TD, which start in 22.02. Over the course of the dialogue, TD is able to articulate his sense of what had happened in the classroom. It is almost as though three people are following their own train of thought, weaving in and out of each other.

Directly following the last comment of TD in 24.40, TC makes a connection and offers an awareness *about* what has just been said:

| 25.33 | TC | That's lovely, and it's a much more enabling sorting out, because if we just sort it out by answering the issue then the next time pupils come up with this issue they have no; they're in no better position to decide; the only resource they've got is to ask Teacher A. But if you sort it out by making them aware this is an issue and making them aware there are consequences for each one and whatever, that is offering them a tool for next time they get in to that situation; so, yeah, I love that. |

I personally had a palpable sense of connection and excitement at the moment above, which I can still recall. This was an awareness made in the moment. It is also a metacomment which, as in a classroom, I conjecture is important in establishing and sustaining a departmental culture of reflection on practice.

CONCLUSION

In both classroom and teacher meeting, there is evidence of established practices that seem to support the noticing of the new, or the gaining of awareness.

In both discussions there is a lack of negative evaluations. If I respond evaluatively, I am explicitly fitting what has just been said into my existing set of beliefs. The possibility for change, or the noticing of something strange only arises with a non-judgemental orientation. Mason (2002, p. 95) writes of the need for developing sensitivities in order to prepare oneself to notice the new; I take it that not being tied in to the habit of evaluating what is said in a discussion is a necessary pre-condition for such sensitising work to take place.

It seems significant that the generalisations being articulated in both discussions followed a period of information gathering, or accounting 'of' (i.e., trying the process on lots of different numbers for the students, and reconstructing the video for the teachers – themselves potentially aesthetic experiences). There is evidence here for the importance of the practice described in Mason (1985) of staying with 'accounts of' before moving to 'accounts for'.

Coming back to Bruner's (1990) question: 'how does what one does reveal what one thinks and believes', the behaviour of the chairs of each discussion (i.e., the teacher in the classroom and TC in the teacher meeting) reveals several things. The fact that comments are not evaluated suggests that the attention of the chair is not on the correctness or otherwise of what is said but more, perhaps, on the internal coherence of what is said, or the relation to some idea (in the classroom this idea was of 'thinking mathematically'). In sharing an awareness *about* what participants have said (i.e., metacommenting) the chairs show they have some criteria against which they hear comments. The listening in the discussions, although non-evaluative, is not such that "anything goes"; certain utterances are privileged in being highlighted and reflected back to the whole group, it seems as though the chairs of the discussion have 'primed or sensitised' (Mason, 2002, p. 66) themselves to notice certain features of discussion which they see as key to their work, and when these comments arise from the group the chairs comment *about* them, 'identifying and labelling' (Mason, 2002, p. 95) what is valued, what is available to be noticed, to the rest of the class. The chairs also show themselves to be open to making connections, and so model a way of being in which one is vulnerable to what is said.

The conclusions above, suggesting the importance of; a non-evaluative stance, metacommenting, 'accounts of' preceding 'accounts for', priming/sensitising myself as a teacher and identifying/labelling in discussion – are all open to testing in the practice of other teachers. Any 'scientific' nature to these findings however will undoubtedly fall short of a causal or necessary connection. Responding without evaluating cannot *cause* others to make connections; no list of actions for a teacher to perform will ensure that participants gain a new awareness or make a connection. Having an aesthetic experience is a matter no procedure can guarantee. Taking an aesthetic approach to education research implies respecting such complexity and uniqueness at the heart of each teaching encounter, whilst not giving up the hope that, through a disciplined noticing, there may be significant and shared awarenesses relevant to working with humans in a manner that fosters creativity, learning and transformation.

REFERENCES

Brown, L., & Coles, A. (1999). Needing to use algebra – a case study. In O. Zaslavsky (Ed.), *Proceedings of the 23rd annual conference of the international group for the psychology of mathematics education* (Vol. 2, pp. 153–160). Israel: Technion (Israel University of Technology).

Brown, L., & Coles, A. (2008). *Hearing silence: Steps to teaching mathematics/learning to teach mathematics.* Cambridge, UK: Black Apollo Press.

Bruner, J. (1990). *Acts of meaning.* London: Harvard University Press.

Coles, A. (2001). Listening: A case study of teacher change. In M. Heuvel-Panhuizen (Ed.), *Proceedings of the 25th annual conference of the international group for the psychology of mathematics education* (Vol. 2, pp. 281–287). Utrecht, NL: Utrecht University.

Csikszenhmihalyi, M. (1997). *Finding flow: The psychology of engagement with everyday life.* New York: Perseus.

Davis, B. (1996). *Teaching mathematics: Toward a sound alternative.* London: Garland.

Dewey, J. (1934–2005 ed.). *Art as experiencee.* New York: Perigee.

Dewey, J. (1929–2008 ed.). *The sources of a science of education.* UK: Lightening Source.

Gattegno, C. (1987). *The science of education.* Reading: Education Solutions.

Heidegger, M. (1962). *Being and time.* Oxford, UK: Blackwell Publishers.

Mason, J. (1985). *Secondary mathematics: Classroom practice.* Milton Keynes, UK: Open University Press.

Mason, J. (1996). *Personal enquiry: Moving from concern towards research, researching mathematics classrooms.* Hampshire, UK: The Open University Press.

Mason, J. (2002). *Researching your own practice: The discipline of noticing.* London: RoutledgeFalmer.

Thera, N. (1996). *The heart of Buddhist meditation.* Sri Lanka: Buddhist Publication Society.

Alf Coles
University of Bristol (United Kingdom)

LAURINDA BROWN

12. SPIRITUALITY AND STUDENT-GENERATED EXAMPLES

Shaping Teaching to Make Space for Learning Mathematics

– Maths is boring.
– I can't do this. I've never been good at mathematics.

Students in classrooms who say these things are caught up in stories of their past experiences of learning mathematics. Mathematics to them is meaningless. What might these students be doing differently?

In this chapter, I explore a range of possible alternative actions for students in their learning of mathematics and how teaching can be shaped to support the students acting. In exploring the contributions of Caleb Gattegno, John Mason and Francisco Varela to links between action and learning, I was struck by the spirituality of the three men and the way that this had informed their awarenesses as teachers and learners. Before describing two extended case studies I will say something about my understanding of spirituality.

SPIRITUALITY

Firstly, for me, spirituality has little to do with any higher being or mysticism. Reading across Eastern and Western religious writings it is hard not to be struck by a common experience that seems to be at the core. Thomas Merton, a monk in an order that lived in silence, was also a prolific writer and writes, in a poem called *In silence*, 'The whole world is secretly on fire'. There is a simple exercise, akin to Buddhist meditation, where you can experience a version of this link between silence and the energetic present:

> Close your eyes, wherever you are currently seated, and listen to the silence beyond the furthest sound. Pay attention first to what is nearest you – I am currently typing this in a coffee house and the steamer is working but behind that I can hear footsteps of people some way away walking around the bookshop and, the noise of the traffic outside in the street ... Keep going ...

I have shared this simple exercise in a number of contexts and many people report that, on opening their eyes at my request, colours are stronger, there is a glow to the world that is not usually, in their day-to-day experience, there. I notice that many people smile broadly as they open their eyes and are there, present in the world, with no thought of tomorrow nor worry about yesterday. It does not need to

S. Lerman and B. Davis (eds.), Mathematical Action & Structures of Noticing: Studies on John Mason's Contribution to Mathematics Education, 147–160.
© *2009 Sense Publishers. All rights reserved.*

last long this state to be relaxing and refreshing. Other people who report not seeing this brightness, found it hard to stay with the task because of various problems or worries. The grace of being in the moment, the energetic present, is spirituality to me.

GATTEGNO, MASON AND VARELA

So, what has this got to do with shaping teaching to make space for learning? If students' attention is in the present they will not be irritably worrying about how they did in mathematics previously, nor being bored.

Gattegno (1970) discusses what he calls the *powers of children* (extraction, making transformations, handling abstractions and stressing and ignoring), *functionings* that can be used in the teaching of mathematics, such as students noticing differences and assimilating similarities; using their power of imagery through asking them to shut their eyes and respond with mental images to verbal statements; generalising given that 'algebra is an attribute, a fundamental power, of the mind' (adapted from pp. 7–11 and 24–27). In Gattegno's writing there is also a model of 'the self', which exists in the present moment, integrating the past (the psyche) and carrying purpose and energy into the future (affectivity).

You can explain all you like as a teacher but there needs to be a connection, a question for the students that your words are answering, for learning to take place. Gattegno would describe this as the 'self at the helm'. The powers of children give a way of planning lessons so that the students' questions, arising from their observations, are at the heart of the lesson. Students are asked to make distinctions, to manipulate objects such as Cuisenaire rods or geoboards and to share their learnings. In turn, these observations and actions are heard and seen by the teacher who is themselves present in their observations, learning about their students and bringing all their past experiences to the task, in this moment, of contingently offering something, such as another action or observation, to challenge the students based upon these perceptions.

Varela, himself a Buddhist and a neuroscientist, in a book *Ethical Know-How: Action, Wisdom, and Cognition*, stresses that "sensory and motor processes, perception and action, are fundamentally linked in lived cognition' (1999, p. 12). For instance, a wine-taster makes ever finer distinctions in tasting wine. Through our repeated actions in the world, what is built up within us that allows us to 'see' (perceive) more? Varela's description of this is in two parts: the first that "perception consists of perceptually guided action"; and the second that "cognitive structures emerge from the recurrent sensorimotor patterns that enable action to be perceptually guided". We 'see' a chair as a sitting-on object. We are, literally, what we do. Learning is seen through effective behaviours in the world, or "immediate coping" (Varela, 1999, p. 5). There is this link between 'seeing' (perceiving) and acting now, in this moment, immediately.

Given the awareness that such distinction-making is at the root of all learning, how is it possible to shape teaching to make space for that natural learning instead of, as Gattegno would describe it, basing our teaching strategies on students memorising. I remember being surprised at the start of my teaching career that students did not start this new lesson from where we finished the last lesson. 'Don't you remember?' I would say. They felt powerless because they could not remember. So, in the style of Gattegno, 'if the students can't remember do something else!'

Mason's 'Discipline of Noticing' focuses attention on us making connections with what we 'see', separating the 'accounts of' the incidents from 'accounting for' them. There is a power in getting students to stay with reconstructing an image or a film. In so doing, questions arise and, through accounting for these observations, students can be led, through stressings and ignorings, conjectures and refutations, to convincing arguments for their own observations.

John Mason was influenced by J.G. Bennett, a student of Gurdjieff. I am more accustomed to the writings of Ouspensky, another disciple of Gurdjieff. My interest in learning sparked my noticing of a discussion in *In search of the miraculous* (Ouspensky). I was reading in comparative religions and was struck by the following description of 'the fourth way', a synthesis of the three, originally seen as separate, "known ways to immortality":

"The way of the fakir:
- through struggle with the physical body
- the teacher serves as an example
- the pupil's work consists in imitating the teacher

The way of the monk:
- through struggle with feelings (emotions)
- teacher as authority figure
- the pupil's work is to have absolute faith in the teacher, in submitting to him absolutely, *in obedience*

The way of the yogi:
- the way of knowledge and the mind
- teacher essential as example and authority but
- the pupil's work is to learn the teacher's methods and gradually learn to apply them to himself." (distilled from Ouspensky, 1977, pp. 44–47)

Combining the other three, "the method of the fourth way consists of doing something in one [sphere] and simultaneously doing something corresponding to it in the other two [spheres]" (p. 49).

Mason has also been explicit, in a striking parallel to the three paths to immortality, about including behaviours, emotions and knowledge in any work on teaching and learning. He extends the aphorism 'only awareness is educable' of Gattegno to:

Only behaviour is trainable
Only emotion is harnessable
Only awareness is educable.

The link with the teaching and learning of mathematics is apparent to me from the writing of Ouspensky in the demand for understanding, "[a] man must do nothing that he does not understand [...] The more a man understands what he is doing, the greater will be the results of his efforts [...] The results of the work are in proportion to the consciousness of the work [...] On the fourth way a man must satisfy himself of the truth of what he is told. And until he is satisfied he must do nothing." (p. 49)

So, seeing these statements as a set of injunctions to a mathematics teacher, shaping teaching to make space for learning involves working with the three aspects of behaviour (doings), emotions (feelings) and awareness (knowings) at the same time, consciously, in an environment where the pupils are also concerned with understanding. This is the path of the fourth way. Another way of thinking about this is that teacher and students are present in the moment integrating their past experiences through their actions and perceptions to develop awarenesses – what does this look like in practice? After one extended case study within mathematics teaching I will explore another account of my practice as a teacher educator, highlighting the parallels.

CASE 1: MATRICES AND TRANSFORMATIONS

What I have spent much of my research life doing since moving to teacher education is what Ruthven (2001) calls 'practical theorising'. He refers to an article of mine (Brown, 1991), where "the main value of this article for our students [student teachers] is that Brown [carefully describes] a sequence of lessons in which pupils tackled a topic investigatively; trying, in the course of this account to answer many of the questions that trouble teachers: how to launch an investigation; when to intervene in pupils' work, and how; how much time to set aside for different stages; how to cope with pupil absences; how to encourage pupils to share their findings; how to help pupils organise and summarise their work; how to draw the different paths of an investigation together and to a close' (p. 172). The article describes teaching through supporting 14-year-old students to explore what I describe as the 'space' of matrices and transformations.

Drawing on the stages of the lesson as discussed in the original article I will reflect in a different way on the teaching strategies used, showing how they are supporting the students to use their powers, as Gattegno has described and how this is also working with the students' and the teacher's present energies taking care of doings, feelings and knowings, the teacher working contingently with the questions and statements arising from the students. These questions and statements are student generated, so the 'examples' that the students work with are from their own awarenesses. Another way of thinking about this is that the students are 'mathematising' – they are doing the process, the 'ing' of mathematics. One awareness, that learning takes place over time, leads to at least a fortnight of lessons being used for any 'space', and often substantially more time is used, the actual time dependent upon the engagement of the students.

So that you have some idea of the activity, what follows is a write-up that is part of the scheme of work of Kingsfield School in South Gloucestershire, UK, where another contributor to this book, Alf Coles, was Head of Mathematics until recently. For the in-depth write-up of the original sequence of lessons, see Brown (1991). After the description of the sequence of lessons I will reflect on the decision-making for the original plan and links to spirituality and student-generated examples.

A first lesson:

Draw a co-ordinate grid in your books from 0 to 12 on x and y ...

Draw one on the board as you say this, or have one already drawn.

... and draw this shape at exactly the same points.

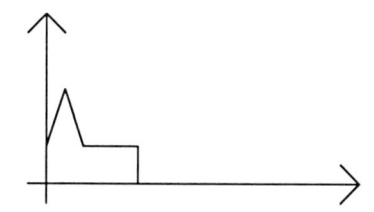

What is it? A church, yes!

We are going to do a process involving these numbers that will change the shape.

We will do it point by point. But first, so we know what we are talking about we need to label each point. Call this point A, what are its co-ordinates? Okay, in this project we write co-ordinates vertically. We are going to work together until everyone can do this process. Copy this into your books as we do it.

1st *row, 1st column, multiply in pairs and add*

2nd *row, 1st column, multiply in pairs and add*

$$\begin{pmatrix} 2 & 1 \\ 0 & 1 \end{pmatrix} \overset{A}{\begin{pmatrix} 1 \\ 5 \end{pmatrix}} = \begin{pmatrix} 2 \times 1 + 1 \times 5 \\ 0 \times 1 + 1 \times 5 \end{pmatrix} = \overset{A'}{\begin{pmatrix} 7 \\ 5 \end{pmatrix}}$$

Repeat these instructions for every point. Invite volunteers to complete.

Label and draw each of the new points:

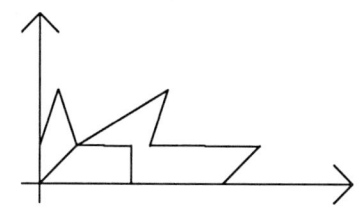

What has happened to the shape?

You may want to introduce the language of stretches or shears and write up things like 'What happens to the area?' if students mention that kind of thing.

Try out your own numbers instead of $\begin{pmatrix} 2 & 1 \\ 0 & 1 \end{pmatrix}$ and see what happens to your own shape.

You may want to discuss sensible shapes to try.

Challenge: *In 5 lessons time I will come in and write a set of four numbers (a matrix) on the board and your task will be to tell me what effect it will have just by looking at it.*

Where this can go

As with 'Functions and Graphs' you may need to share ideas about how to explore this challenge in an organised way. With weaker students the investigation can be limited to matrices of the form

$$\begin{pmatrix} a & 0 \\ 0 & b \end{pmatrix}$$

If students work on paper this can be pinned to a board. They must write what the matrix has done to their shape before they pin the paper to the board. Questions may arise from looking at the matrices others have tried. Someone could organise the class results and feedback how they did it. *What do other people notice? So, what questions could we ask? Can you generalise or predict? What happens to the areas of your shapes?*

At some point focus the class on the basic transformations: reflection, rotation, enlargement, stretch, identity, crush (ie when the shape becomes a line or point), shear. Ask if anyone has a matrix that does one of these things only. Create a space on the board for students to write up matrices that do these basic transformations, see if anyone can generalise – e.g. is $\begin{pmatrix} n & 0 \\ 0 & n \end{pmatrix}$ an enlargement scale factor n.

Someone could organise the class results under these headings.

Some students may even want to extend such generalisations – e.g., to 3-D.

Get students to try and find the inverse matrix – e.g., of an enlargement.

REFLECTIONS

When I am teaching I expect my students to act. For instance, I have used the offering of at least two different but similar images for students to describe what they see across many different activities. In supporting students to act I know, from experience of asking 'what's the same, what's different', that they can use their powers of discrimination. What I then do is contingent upon what is said. The students learn the mathematics whilst I learn about the students' learning. In this

SPIRITUALITY AND STUDENT-GENERATED EXAMPLES

case, for the students to be able to explore the space of matrices and transformations they need to be able to cope with the procedure. I want a shape that is simple enough to have a few points but brings to mind an iconic shape, so that it will be recognisable after transformation. So, I am asking students to watch the difference between the two images before and after the application of the matrix to the original coordinates of the shape.

From the original writing some insight into these choices is given:

> 'We go through the process for each of the points, with me exploring who can do the process and who can't. At one point, I ask, 'Is there anyone who could not do this one?' No one puts a hand up and, watching carefully, I believe them. If someone had, I would have worked with them. As it was, we finish off the points with me choosing people at random to tell me what to do and we stare at the new, different, church. I had to make a decision – do we discuss what this shape looks like or do I offer them a challenge and set up their task?' (pp. 7–8)

I am struck reading this about the watchfulness described both of me watching the students' reactions to my question and of me acting as a model of watchfulness as 'we stare at the new, different, church'. We are all in the present moment, staying with what we see. I then give the group a chance to check out that they can do what is asked using their own, self-generated task and also give them a challenge that will allow them to develop their ideas.

In the original chapter the challenge that was given read:

> 'In five lessons' time, I will come in, write any old matrix on the board and expect you to be able to tell me what effect it would have on a shape without actually drawing it out.'

I then wrote:

> 'There were excited intakes of breath around the room: they felt challenged and were going to engage. I felt relief.'

I knew what I was looking for, the energetic present. These students now had an intention, the affectivity into the future that would allow them to make decisions about what matrix to try next and why. I could relax because the feeling of this group who were new to me was the same as previous groups. I set them the first task to support them in checking out that they could do the procedure and also give the students a whole range of examples from the space to compare and contrast, using their powers of discrimination again.

> 'For a beginning, and to check that each of you can do a basic transformation, I'd like you each to take a piece of this paper, choose your own shape and matrix, and see what happens. When you have finished, we can put them on the display board over there or on the ceiling. Then you will have a collection of a lot of information from which to work.'

My attention focused on each student being able to do the process. Their attention was on creating the display, which happened extremely quickly.

When I have found a 'space' I use it again and again, continually being surprised by new questions raised by students I work with. As I work, my own attention and awarenesses are educated about the possibilities of the space and more of my focus can be on learning what this new group of students does – being aware of the samenesses and differences between what has happened before and what is happening now. Each new experience of using a 'space' can feel like the students are doing more, whereas, in fact, what is happening is that as a teacher I am able to see more each time in what students are doing. It's like a mechanism for slowing down experience because more distinctions are automatic. I used to call such packed awareness in observation 'slow-space' and would talk mysteriously about needing to be in slow-space when teaching. I now recognise another aspect of spirituality in this account, in the present moment time is slowed down since we are not overtaken by busy distractions where present time can disappear very quickly.

What is it I see? A certain diversity or range of strategies being used? Certainly difference. All I know is that allowing myself to notice or learn about my students leads to instant decision-making – I am doing something – the immediate coping of effective behaviour.

Past experiences using the space inform my sense of fruitful ways to go:
- Which matrices do the usual transformations for us?
- Is there a link between the matrix and the area change?
- There is a lot of information at the moment, which perhaps feels confusing. What about ways of simplifying our task – by getting organised? Simpler shapes? Being organised about which matrices to choose?

But if I follow my own intuitions and offer them then this can limit the students doing the mathematics. I offer in relation to the 'ing', the process, and leave the mathematics to the students.

I like the students to use a board in the classroom to share their conjectures or thoughts and others can come and talk with them about these. This is another take on student-generated examples – more sharing of the culture of mathematics. Here are a few comments from the 'Questions and Comments' board (Brown, 1991):
- What's the simplest shape I can make? Square? Triangle?
- Will the same two numbers on the top and bottom turn out in a squash?
- With a triangle you always end up with another triangle but in a different size.
- These two numbers will make enlargements.
- Will the one in the top left corner and a zero in the other top place keep the x-coordinate the same?

These are real questions for the students. Some, like the first, I would not ever have thought of before it was asked, others feel natural in that they occurred when I worked on the problem. The basic challenge leads to the students trying to find what each number in the 2×2 matrix 'does' separately. I have known base vectors to emerge. However, the point of all this is not so that the year 9s in the chapter or the year 8s in the lesson write-up above, end up being able to do matrices and transformations, it is so that they explore the space and, in so doing, explore area, coordinates, negative numbers, shape, transformations, all within a space where they are dictating what to explore.

At the end of the project, the following were the focus questions that the students did not feel had been answered: 'Exploration of all matrices made up of 0, 1 and −1; What about three-dimensional transformations?; Inverse matrices – or what would I use to take the image back to the shape I started with?' (Brown, 1991, p. 12) It was towards the end of the summer term and some students carried their work on over the holidays.

Many of the students had the energy that some authors relate to creativity - absorption and staying power. They had been using all their powers of discrimination, abstraction (some used algebraic representations to support their generalisations) and transformation. Such activities as classification, organising, generating examples, exploration and a start based on using powers of observation to see difference can be the base for planning all lessons.

A CONTEMPLATIVE PRACTICE

O'Reilly (1998) talks about the same ideas with a different language. In a chapter called 'To teach is to create a space' a phrase attributed to Parker Palmer, she says:

> 'Most of us believe, at some level, that what happens in the classroom is caused by the teacher. In reality, we cause or control very little. To "create a space" acknowledges both our sphere of responsibility and our lack of control' (p. 2)

This resonates with Gadamer's thought:

> 'The way one word follows another, with the conversation taking its own twists and reaching its own conclusions, may well be conducted in some way, but the partners conversing are far less the leaders of it than the led. No-one knows in advance what will 'come out' of a conversation.' (1989, p. 383)

Developing conversations within classrooms where the teacher is being led by the conversations involves 'keeping quiet much more that we have ever imagined possible, and ... listening more astutely than we have before' (O'Reilly, 1998, p. 2).

THE NOT-YET-IMAGINABLE

Davis and Sumara (2007) 'use the term *not-yet-imaginable* to refer to that space of possibilities that is opened up through the exploration of the current space of the possible' (p. 58) and this phrasing resonates for me with the process of working in the spaces I have described above. What happens in the exploration of the mathematical spaces I work in with students is different each time and I am not attached to any prediction of what will be the grit in the oyster emergent through the interactions within the members of the group. Working on the mathematics within the spaces myself before entering the interactive space helps to attune my awarenesses but then all this is let go as my attention turns to learning the students working in the space. This orientation also serves to support students 'toward unrealized prospects' (p. 58), given that they are not limited by the explicit content expectations of the teacher.

DELICATE SHIFTS OF ATTENTION

Mason writes about shifts of attention leading to abstraction in mathematical learning (e.g., Mason, 1989). These shifts of attention seem to me to be part of what teachers do in opening up the spaces of the possible. Lending our attention to those who aren't using it themselves (O'Reilly). What does the teacher comment on in these spaces as they learn about their students' learning?

In establishing a classroom culture at the start of their first year in a new school, Alf Coles, a classroom teacher and research collaborator (Brown & Coles, 2008), uses the idea of what we came to call 'metacommenting' (from 'metacommunication', Bateson, 2000) to focus the attention of his students on behaviours that will support their 'becoming mathematicians' or 'thinking mathematically'. A working learning culture is developed within the group that is somewhat different each year given that the range of what students do and notice is different.

When students are engaged in an activity, Alf comments on behaviour that he notices as being mathematical. Students write at the end of an activity (usually at least six lessons) about what they have learnt both at the skills level and about thinking mathematically. Typically the students work with the language of 'conjecture', 'proof', 'theorem', 'testing conjectures' and 'counterexample' because this is the language used by their teacher. The dialogue below is recorded in my notes from a lesson. Metacomments are in 'italics'; Alf speaking is indicated by '–' and students by '~' at the start of lines:

– There were some counterexamples to that. Remind me what that is.
~ One that does not fit the conjecture.
– OK, Ben has done something very mathematical. He's *gone back and looked again and changed it* [the conjecture].
~ [Later in the same lesson.] All two digit numbers will add up to 99. [David's conjecture is written on the board.]
~ I've got another counterexample to Ben's.
– This is how mathematicians work; *are there counterexamples? Are two conjectures actually linked* and so on.

Other metacomments that Alf has used are 'getting organised', 'asking and answering your own questions' and 'listening and responding' (to what other students say). So, because Alf is talking about what students are doing that fits with 'thinking mathematically' some of these metacomments capture the attention of the students and become their purposes, giving them a sense of knowing what to do. Over time they develop strategies to work on problems and learn skills.

For myself, working with student teachers, I comment on and notice when they in turn metacomment in and on their own classrooms – I call this their 'patter' – the ways in which other people develop a sense of how they want others to behave. It does not seem to matter at many levels what that patter is, as long as it is something that the student teacher cares about. What seems important is that the patter exists! The students or student teachers learn what they need to and the teacher or the teacher educator is learning about learners' strategies and behaviours as the culture develops in the group.

In their article, Davis and Sumara argue for a reconceptualisation of the teacher's role in learning away from a pre-planned agenda and they also give examples from teacher education and university teaching. What follows, after the example above of myself as a mathematics teacher without a pre-planned agenda, is an example of how this works in my practice as a teacher educator.

CASE 2: TEACHING HYPERBOLIC FUNCTIONS

For two days, at the end of their one-year course, my student teachers of mathematics teach each other topics from A-level mathematics identified through discussions between themselves. Hyperbolic functions had been suggested as one area never visited by some or felt to be in need of more work by others.

What arose for me were memories of working on hyperbolic functions with a group of students who would go on to university to read mathematics or mathematics-related subjects, the 'further mathematicians', near the start of my life as a school teacher. The chapter in the text book was full of identities and properties that took, so other members of staff reported, many weeks to cover. I offered an entry into a 'space'.

> You are a mathematician working for a company and are asked by the resident physicist to write a report on two functions that keep appearing in a research project. The two functions are:
>
> $$f: x \to (e^x + e^{-x})/2 \qquad g: x \to (e^x - e^{-x})/2$$

I mentioned this starter to the group of student teachers who were going to be teaching hyperbolic functions, when they were thinking about what to do in the session. The group responded to my starter and worked at adding, subtracting, multiplying, dividing, squaring, sketching, differentiating ... the two functions. "Well, that's your planning done. You won't need to do anything else", I said.

When the time came for the group to teach their hour session, I felt dissonance when, as they began, the integral of $1/\sqrt{(x^2 - 1)}$ was written on the board accompanied by the comment "I want to integrate this and the only things I have at my disposal are $f(x)$ and $g(x)$." I could see that this offer of the integral came from a question that had arisen for the teaching group of "I wonder why you might need to use hyperbolic functions anyway?" This seems like a useful question for them in preparing the session, but what would it be like as a starter for working mathematically?

One of the student teachers who was leading the session came up to me after it was over and commented that he was surprised that no-one had sketched the graphs of the functions (he had done this in response to my question). I noticed, from looking at the work myself and another student teacher had done on the task, that we had not sketched the curves in response to the starting task either.

We linked the periodicity of the function as we differentiated with $\sin(x)$ and $\cos(x)$. We also made the connection with the ways in which the trigonometric functions are used in substituting to simplify integrals. We, therefore, checked whether there was an equivalent for $\sin^2(x) + \cos^2(x) = 1$ by squaring each of f(x) and g(x) and finding the difference between the two new functions was 1. We then substituted in the integral ... and stopped working.

Thinking about the story: in working with groups of further mathematicians used to exploring within mathematics lessons throughout their school life, there was no explicit goal and, although the very beginning of the lesson felt similar, the students did not stop working. Sketches of the functions were made. At the point at which a similar relationship was found to $\sin^2(x) + \cos^2(x)$, all the trigonometric identities were shared out and other questions emerged from the students, which provided motivation for further work, such as, "It must be possible to find a defining relationship between the two systems, the new one and the trigonometric functions". The offer of the task created a divergent space. The chapter content was covered in a lesson!

I learn best through doing. I had been thinking about spaces in preparation for writing this chapter and, from this experience, I could make a distinction between a 'space' compared to what I might call problem-solving. Instead of expanding the space of the possible, there seems to be nothing to do when you've solved the problem, except to feel pleased with yourself. The student teacher's role as teacher became supporting links between what was being explored and the solution of the problem as set. The work in the space is what might be called divergent and for problem-solving convergent. However, the space of the group's original explorations with me when planning to teach raised the awareness for them of the difference between problem-solving and when they had worked on the original challenge themselves.

I offered my reflections about what was happening, comments at a meta-level to their actions, to the student teachers. In the discussions that followed my observations, there were reports from student teachers who were still working on a range of questions that had arisen for them in tackling the problem. Not everyone had stopped working having solved the integral! There is, of course, nothing to stop a convergent problem becoming an entry into a space. For example, consider the statement, 'Find me a square of area 10' (the corners of the squares must lie on dots of a square grid). The teacher's role, here, is to work with the students' questions, such as 'What numbers are not possible as areas of squares on dotty grid paper?' as they arise e.g., 'Let's make a collection of areas we can do', 'Is there a quick way of finding the areas of the squares?' Answering one question still leaves unresolved issues. Reflection on practices in the moment of teaching leads to awarenesses that can be used in future practice. For example, there is now the possibility, for that student teacher, of response to myself and the other student teacher sitting feeling pleased with ourselves having 'finished' of something like 'And what are you going to do now?' There can then be an awareness of creating a culture in the classroom of learners looking to extend their thinking beyond initial questions.

So, I am clear that the student teachers will be able to use their powers of discrimination to learn about teaching mathematics and that it is important to allow *them* to find teaching strategies that work to support the learning of their students. The fact that that is how I expect us to work needs setting up. I explicitly state that I do not have an image of the best way to teach. At the start of the course, the first time the student teachers have a session with their mathematics education tutors at the University, they have already spent two weeks observing and working in a primary classroom. The students are invited to bring to mind one image that seems to them to be about 'teaching' arising from their recent experiences and one that seems to be about 'learning'. Based on these real experiences they must be prepared to share with others in the group completed slogans for 'teaching is ...' and 'learning is ...'. This activity serves to orient the students to see the 'shape' of their classroom interactions and allows issues to arise that might become purposes for them at a later time. The activity serves to stop them from living at the level of reporting long held beliefs or attitudes divorced from their recent experiences. For me, however, 'gaining a sense of the teacher they want to become' is a purpose, since I have a range of behaviours as a teacher educator related to offering this task and I recognise their purposes as they are articulated or enacted and can metacomment on the range of those.

In 1995, John Mason and I gave a pair of lectures at the Association of Teachers of Mathematics (ATM, one of the professional associations for teachers of mathematics in the UK) Easter Course in the UK under the heading *Shaping teaching to make space for learning*. John gave the opening lecture, focusing on the mathematics and illustrating how teaching develops (Mason, 1995) and I gave the closing lecture, focusing on teaching and illustrating how mathematics develops (Brown, 1995). In the collection of materials left over from the ATM lecture were the unused slides of the three paths to immortality and the fourth way. I had intended to be explicit about the spirituality inherent in my ways of working back then but, in the end, had left this unsaid. John Mason explicitly states: "I see working on education not in terms of an edifice of knowledge, adding new theorems to old, but rather as a journey of self discovery and development in which what others have learned has to be re-experienced by each traveller, re-learned, re-integrated and re-expressed in each generation ... each effective act of teaching is an action which involves the mutuality of three impulses" (p. 177). It is this aspect of staying with the details of perception, within an activity energetic with intention, and using and developing new awarenesses and using skills, that is the fourth way – allowing such noticing involves us in being present, now, not distracted by the noise of our emotions, worries or joys, in other words being here and now, open to our awarenesses.

> 'And he is not likely to know what is to be done unless he lives in what is not merely the present, but the present moment of the past, unless he is conscious, not of what is dead, but of what is already living.' (Eliot, 1922)

REFERENCES

Bateson, G. (2000). *Steps to an ecology of mind*. Chicago, IL: University of Chicago Press. (1st ed., Paladin, 1972).

Brown, L. (1991). Stewing in your own juice. In D. Pimm & E. Love (Eds.), *Teaching and learning school mathematics* (pp. 3–15). London: Hodder and Stoughton.

Brown, L. (1995). Shaping teaching to make space for learning. *Mathematics Teaching, 152*, 14–22.

Brown, L., & Coles, A. (2008). *Hearing silence*. Cambridge, UK: Black Apollo Press.

Davis, B., & Sumara, D. (2007). Complexity science and education: Reconceptualizing the teacher's role in learning. *Interchange, 38*(1), 53–67.

Eliot, T. S. (1922). *The sacred wood*. London: Methune.

Gadamer, H. G. (1989). *Truth and method*. London: Sheed and Ward.

Gattegno, C. (1970). *What we owe children: The subordination of teaching to learning*. London: Routledge & Kegan Paul.

Mason, J. (1989). Mathematical abstraction as the result of a delicate shift of attention. *For the Learning of Mathematics, 9*(2), 2–8.

Mason, J. (1994). Researching from the inside in mathematics education: locating an I-you relationship. In J. da Ponte & J. Matos (Eds.), *Proceedings of the 18th annual conference of the international group for the psychology of mathematics education* (Vol. 1, pp. 176–194). Lisbon, PT.

Mason, J. (1995). Shaping up. *Mathematics Teaching, 152*, 4–11.

O'Reilly, M. (1998). *Radical presence: Teaching as a contemplative activity*. Portsmouth, NH: Heinemann.

Ouspensky, P. D. (1977/1949). *In search of the miraculous: Fragments of an unknown teaching*. Orlando, FL: Harcourt.

Ruthven, K. (2001). Mathematics teaching, teacher education, and educational research: Developing "practical theorising" in initial teacher education. In F. -L. Lin & T. J. Cooney (Eds.), *Making sense of mathematics teacher education* (pp. 165–183). Dordrecht, NL: Kluwer.

Varela, F. J. (1999). *Ethical know-how: Action, wisdom, and cognition*. Stanford, CA: Stanford University Press.

Laurinda Brown
University of Bristol (United Kingdom)

Mathematical
Action &
Structures
Of
Noticing

Part 3:

VARIATION AND MATHEMATICAL STRUCTURE

Part 3:

VARIATION AND MATHEMATICAL STRUCTURE

As developed in Part 2, the discipline of noticing has been a core theme throughout John's work. One of the ways that this emphasis has been articulated is through the role of variation in perception and learning – that is, the sharpening of one's awareness that comes about by focusing on what changes and what stays the same when one engages in mathematical activity, or many other ways of being in the world for that matter.

To really appreciate this point, it is helpful to know that, for John, 'awareness' is not a singular phenomenon or static ability. Three forms discussed in his work are *awareness-in-action* (the ability to act in the moment), *awareness-in-discipline* (awareness of awareness-in-action), and *awareness-in-council* (awareness of awareness-in-discipline – or, awareness of awareness of the ability to act in the moment).

In this frame, teaching mathematics all about educating awareness(es), which in John's work has been articulated in the inextricably entangled themes of the role of variation in perception and the structure of mathematics tasks. In terms of teaching, it is all about the choices that a teacher makes – or can make – to frame tasks that point students towards attending to features of mathematics that are distinct or can be seen to be the same. These are aspects of the structure of mathematics, whose nature and presence have to be made available to the awareness(es) of learners, including teachers-as-learners, for an appreciation and enjoyment of mathematics to occur.

Rina Zazkis, Nathalie Sinclair and Peter Liljedahl open the section by applying John's three forms of awareness to document their students' and their own development. They discuss successive stages of an experiment in which student teachers construct 'lesson plays', scripts for lessons that show the interaction of teacher with students, in the process working to enable teachers to see what students are attending to. In the chapter that follows, Pedro Gómez and María José González extend the discussion of awareness by taking up one of John's questions: "Teachers set tasks because they believe that working on tasks will promote learning. But what learning might ensue from a given task, and how can tasks be selected to promote certain kinds of learning?" (Mason, 2004, p. 25). They propose an heuristic procedure for analyzing the cognitive demands of school mathematics tasks.

In the following chapter David Pimm and Nathalie Sinclair look at the elements of the mathematical, the historical and the pedagogical to examine the notion of a culture of generality and, they suggest an associated pedagogy. Finally, Andy Begg focuses on noticing that looks back and takes one forward, in the sense of developing experience, of working on mathematical tasks with students. He reflects on the process of his own development as a doctoral student of John's as part of the reflection backwards.

PART 3

In the final chapter of this book, written by Anne Watson, John's partner in life and work, the most recent twists and turns in his thinking, sometimes alongside and sometimes distinct from Anne's own work, are described. Anne draws attention to the symmetry in the need for the teacher to work on the structuring of students' mathematical attention, which opens awareness of the structures of mathematics.

RINA ZAZKIS, NATHALIE SINCLAIR AND PETER LILJEDAHL

13. LESSON PLAY

A Vehicle for Multiple Shifts of Attention in Teaching

MOTIVATION

As University professors working in mathematics teacher education, the 'methods' courses for prospective teachers are part of our usual teaching assignment. 'Planning for instruction' is one of the topics that are explored through these courses in different settings. It has been the students' expectation that the 'methods' courses will provide them with an opportunity to develop a variety of 'lesson plans.' These plans would have a dual purpose: first, to satisfy some of the requirements of the course, and later to be used as already prepared modules in their teaching.

Traditional lesson planning is based on a linear sequence that starts with a declaration of behavioural objectives and ends with a description of assessment techniques that should determine to what degree the objectives have been met. Despite the contemporary prevalence of student-centred approaches advocated in teacher education, the traditional teacher-centred mode of instructional planning, as taught in teacher education programs, has not changed.

Following prior experience with the 'general', that is, not 'subject matter specific' methods of instruction, many prospective teachers interpret creating plans for a lesson as filling in boxes of predetermined rubrics, such as SWAT (students will be able to), goals and objectives, assessment practices, teacher/student activities and explicit timelines. Having examined a variety of lesson plans created according to standard templates, we felt uneasy. On one hand, we recognized many worthwhile features in the presented plans, such as the use of manipulatives and other teaching aids, appropriate choice and sequence of activities, and different modes of engagement, such as lecture, group work, or individual student work. However, we felt that in most of the lesson plans, even well crafted ones, the main ingredient in preparing for instruction was missing – that of interaction with students and attention to students' difficulties. As such, we sought a vehicle that would direct prospective teachers to what Mason (1998) refers to as "a fundamental question of teaching" (p. 247) – namely, "What are students attending to?" Simultaneously, we were seeking an answer to a related question for teacher education, again, as formulated by Mason, "How can we enable teachers to see what students are attending to?" (ibid.).

With these questions in mind, we suggest a technique for implementation in teacher education, which can either enhance or replace the traditional approach to

S. Lerman and B. Davis (eds.), Mathematical Action & Structures of Noticing: Studies on John Mason's Contribution to Mathematics Education, 165–177.
© *2009 Sense Publishers. All rights reserved.*

planning. We refer to it as *lesson play*, which involves presenting part of a lesson in the form of a play or dialogue, which is set in a classroom, where the characters/players are the teacher and the student(s). In Zazkis, Liljedahl and Sinclair (2009, p. 40) we provided a detailed analysis of the lesson play construct as "a novel juxtaposition to the traditional planning framework as a method of preparing to teach a lesson" and discussed its pedagogical affordances. Here we use Mason's (1998) classification of three different forms of awareness to document our own evolution of the task design, as well as that of the prospective teachers.

> *Awareness-in-action* is the ability to act in the moment. This corresponds to Vergnaud's notion of 'theorem-in-action', the term that describes students' acting according to a certain rule or constraint, without being able to state it explicitly. An individual may not be explicitly conscious of his/her own awareness-in-action, but it can be recognized by others through the individual's actions. A student may demonstrate this kind of awareness by performing an action such as adding fractions, but his skill may be limited to 'knowing how' rather than 'knowing why.' This level of awareness in teaching is recognized when a teacher poses a certain question, corrects a mistake, suggests an answer or selects a task, but is unable to justify or explain his/her choice.
>
> *Awareness-in-discipline* is awareness of awareness-in-action. This awareness is essential in order to articulate awareness-in-action for others. According to Mason, the one important distinction between the two kinds of awareness involves the ability "to do" in contrast with the ability to instruct others. Teachers who possess awareness-in-discipline are able to articulate the choices they make in instructional situations.
>
> *Awareness-in-council* is awareness of awareness-in-discipline. This awareness is essential in order to articulate awareness-in-discipline for others. It describes one's sensitivity to what others require for building or enhancing their awareness.

In what follows, we present the evolution in working with lesson plays in our teaching and research. This evolution represents own developing awareness of how to enable teachers to direct their attention to different aspects of mathematics and pedagogy.

LESSON PLAY – FIRST STEPS

As one of the group assignments in the methods course, we asked prospective teachers to design and write a lesson play – a script for a lesson that described an interaction between a teacher and a group of students. The topic for the lesson was left wide open, with a shared understanding that it should relate to the school curriculum. Prospective teachers could further choose the age group in which their imagined lesson would take place and the size of the group with which the teacher communicated.

Not surprisingly, we say in retrospect, our first steps were doomed to failure. The produced lessons looked like a capture of a lecture with limited interruptions. That is to say, the plays included long monologs by a teacher character, with occasional questions for students. In most cases, students provided correct answers to the posed questions. The teacher character praised them and followed up with further explanations and some additional questions. In a limited number of cases, where an imaginary student suggested an incorrect answer, the teacher immediately addressed the class, asking whether someone had a "different idea," and then a desired solution was put forward in the script.

With this first attempt we felt that the attention of prospective teachers was mainly focused on self-delivery of the content, rather than on their students. In a way, they were acting out a 'standard' lesson plan. Their responses served for us, using Mason's (2002) phrase, as a "form of disturbance which starts things off" (p. 10) that triggered the next phases of development. We needed to re-design the lesson play task in order to help move the prospective teachers from their awareness-in-action, which seemed to be prompting them to re-create familiar interactions with students, to an awareness-in-discipline.

LESSON PLAY – SECOND STEPS
(SHIFT OF ATTENTION TO A POSSIBLE PROBLEMATIC)

In our second attempt to engage prospective teachers in writing lesson plays, in order to foster interaction with students and students' difficulties, we added a requirement that the lesson should address a specific problem that a learner may encounter with the content. Yet again, the choice of the problem, as well as the choice of the content, was left open. In this iteration of the task, prospective teachers were able to identify a problem, but their ways of approaching problems were rather limited. In most scripts, the way to deal with an imagined problem was first to get a correct answer and then to turn "back to the basics," that is, to re-teach the content or restate the 'rule' rather than attend to a specific naïve conception causing the problem.

The following is an illustrative example:

Teacher: Today we'll practice multiplication of decimals. What's 0.2×0.3?
Gary: 0.6
Teacher: Does someone have a different answer?
Annie: 0.06
Teacher: This is correct. Recall that our answer has to have the same number of digits after the decimal as the numbers in question. So in this case our answer will have 2 digits after the decimal.
Susan: So can't it be 0.60?
Teacher: No, because 0.60 is the same as 0.6. Remember, decimals are also fractions. So $^{60}/_{100}$ is the same as $^{6}/_{10}$. So in fractions the answer is $^{6}/_{100}$.
Susan: So can we write the answer as $^{6}/_{100}$.

Teacher: No, since we are working with decimals, we write this as 0.06. Now, let's do a few more examples.

We recognize in this excerpt that the prospective teacher is able to predict a possible confusion, but is not dealing with it in a skilful manner. She explained not only the 'rule,' but also a motivation for the rule by drawing a connection with a common fraction. In terms of shift of attention, we recognize that the teacher's attention is focused on helping students get a correct answer, rather than on attempting to understand and address their difficulties. In terms of awareness, we suggest that the teacher's response indicates that she has not yet developed awareness-in-discipline, and her instructional behaviour is an attempt to transform self-awareness-in-action, that is, ability to perform calculations correctly, to students' awareness-in-action.

Looking at the collection of plays produced by teachers at this phase, we further found that their personal 'example space' of possible mistakes was very limited, and most of the examples of mistakes generated by imaginary students dealt with fractions or incorrect column subtraction. The latter example was previously discussed in class when the topic of deliberation was 'buggy algorithms' (Ashlock, 2005). This choice of topics is hardly surprising as personal example spaces are said to be "triggered by current task, cues, and environment, as well as by recent experience" (Watson & Mason, 2005, p. 76). Yet again, a more specific direction was needed, and it emerged in our next phase.

LESSON PLAY – THIRD STEPS
(SHIFT OF ATTENTION TO A SPECIFIC PROBLEMATIC)

As the third iteration of the task, we engaged prospective teachers in writing and presenting scripts for lesson plays, based on the provided prompts of interaction. Each prompt presented a common error in a student's reasoning. These prompts included the following:

- There are 20–25 students in the classroom, working on the following problem:

  Once upon a time there were two melon farmers; John and Bill. John's farm was 200 m by 600 m and Bill's farm was 100 m by 700 m. Who grew the most melons?

  S/he sees that the student has written:

  They both grew the same amount.

- There is a conversation between the teacher and a student. There are 20–25 students in the classroom.

  T: *Why do you say that 91 is prime?*
  S: *Because it is not in our times table.*

- There are 20–25 students in the classroom, working on the following problem:

A toy train has 100 cars. The first car is red, the second is blue, the third is yellow, the fourth is red, the fifth is blue and sixth is yellow and so on.

(a) What is the colour of the 80th car?
(b) What is the number of the last blue car?

The teacher is moving through the room observing how the students are progressing. S/he stops and points at one student's work.

T: *Why is the 80th car red?*
S: *Because the 10th car is red. So, the 20th car, the 30th car, the 40th car, and so on, will be red.*

- There are 20–25 students in the class, working on the following problem:

A toy train has 100 cars. The first car is red, the second is blue, the third is yellow, the fourth is red, the fifth is blue and sixth is yellow and so on.

(a) What is the colour of the 80th car?
(b) What is the number of the last blue car?

The teacher is moving through the room observing how the students are progressing. S/he stops and points at one student's work.

T: *Why is the 80th car red?*
S: *Because the 4th car is red, and 80 is a multiple of 4.*

The prospective teachers were asked to identify a possible source of the error, consider instructional treatments that attend to the error—without immediately correcting it—and, as in the previous iterations, present those in a dialogue form. These detailed requirements prompted an explicit focus of attention on the students.

The provided prompts were drawn from our 'expert' example space – they included commonly known errors (drawn from conventional example space), such as confusion between perimeter and area, and also difficulties identified in our prior research related to prime numbers and division with remainder (Liljedahl, 2004; Zazkis & Campbell, 1996). They presented prospective teachers with a variety of choices for their assignment, while ensuring that the errors they had to confront were significant enough to prompt thoughtful treatment.

LESSON PLAY – FOURTH STEPS
(SHIFT OF ATTENTION TO LANGUAGE AND ARTICULATION)

The fourth iteration was similar to the third one, but with two important variations. While all the previous lesson plays were designed and written by groups of 3-4 students, this time we asked prospective teachers to submit their plays individually. They could plan and discuss possible approaches to the presented prompts in their groups, but the write-up had to be completed separately. We believed that this would help in developing a personal voice and – though the approach could have been shared by group members – the individual write-up would create a further shift of attention on the language – both imagining and imitating students' language and choosing personal language in a teacher's response. Given the importance of language in mathematical thinking, we wanted every prospective teacher to have to choose the specific words they would use to convey ideas, offer definitions, and respond to student-generated statements. Indeed, when the prospective teachers moved from their group discussions, which generated statements such as "I would just show her ..." or "I'd simply get her to explain what she did ..." to the crafting of their play, they found that articulating those simple strategies was not a straightforward task.

The second variation was that we asked students in this iteration to include a diagnosis (written in regular, paragraph form) for the presented error, that is, their speculations of what could have caused the imaginary student's erroneous response. The diagnosis, which required consideration of how an error had been made, and how it might be taken up, would force a more explicit awareness-in-discipline.

EXCERPTS OF 4th GENERATION LESSON PLAYS

In what follows we present excerpts from several lesson plays and discuss shifts of attention that are present in these plays. We acknowledge that a teacher's attention is directed simultaneously to a variety of issues, and therefore focusing our analysis on one particular shift does not mean ignoring other aspects.

We invite the reader to consider the following excerpt in terms of shifts of attention that various teacher characters undertake (and that, of necessity, what the prospective teachers writing the lesson play have imagined).

Excerpt 1: pedagogical and mathematical shifts of attention

Mrs. Green: Why is 80th car red?
Andrew: Because the 4th care is red and 80 is a multiple of 4.
Mrs. Green: I see, 80 is definitely a multiple of 4. How did you reach your answer?
Andrew: Because we just skip-counted by 4's and we called out 80 so that must mean 80 is a multiple of 4. Since the 4th car is red, the 80th car will be red.
Mrs. Green: I see. So what you're saying is that all multiples of 4 will be red?
Andrew: Yes!
Mrs. Green: OK, Let's test your theory and break it down into smaller numbers. Using the blocks, arrange them into the pattern of red-blue-yellow and find out what colour the 8th block is.

Mrs. Green walks away while Andrew finds the box of blocks and arranges them into the pattern. She circulates among other students while Andrew works on his own. After a few minutes she returns.

Mrs. Green: So Andrew, what did you discover?
Andrew: I put the blocks into patterns but the 8th car isn't red.

[…]
Mrs. Green: Well. So far, what do we know?
Andrew: That the 4th car is red, the 8th car is blue and the 12th car is yellow
Mrs. Green: Right, and these numbers are all multiples of what:
Andrew: 4
Mrs. Green: What are you thinking now: Has anything changed?
Andrew: I don't think the 80th car is red anymore, But I still don't know what colour it is. What do I do?
Mrs. Green: I don't know the answer, I was hoping you would find that for me. Let's take a look at another pattern. What happens if we add another colour, green, to the train? So the pattern is red, blue, yellow, green?

Mrs. Green circulates and returns in a few minutes.

(In what follows Andrew makes a connection between the suggested pattern of 4 colours and a multiples of 4, understands that he has to look at multiples if 3 when considering the original problem, skip counts up to 78 and draws a conclusion about the 80th car, continuing the repeating pattern from 79.)

We recognize a variety of pedagogical shifts of attention in this excerpt. We consider as 'general pedagogy' a teacher's move that is not specific to the task. That is, a move that can be applied in a different situation, not necessarily mathematical. This includes requests to clarify ("How did you reach your answer"), reflection/ rephrasing and extenuation ("what you're saying is that all multiples of 4 will be red?"), directing a student to use manipulatives, putting the student in charge (by suggesting "I don't know the answer, I was hoping you would find that"), and walking away to allow the student to make progress on his own. In this last action, by removing herself from the discussion, the teacher fosters exploration, discovery and self-reliance.

We further recognize shifts of attention that are mathematical in nature and directed towards the specific problem. We consider those as 'mathematical pedagogy' (Mason, 2007). To help Andrew discover his error, Mrs. Green suggests to "break it down into smaller numbers." This may be an implicit enacting of Pólya's powerful problem-solving strategy: think of a simpler but similar problem (Pólya, 1945/1988). While Mrs. Green explicitly instructed the student to consider the 8th car, in other excerpts it was at times left open for the student to decide which smaller number to focus on. In other lesson plays we found the teacher character skilfully leaving the specific but a-priori choice for a student, asking "give me a multiple of 4 under 25," in order to continue the exploration with the student's choice.

Moreover, attending to mathematics and to Andrew's mathematics, Mrs. Green's intention is not only for Andrew to discover that considering multiples of 4 is unhelpful, but also to exemplify in what case looking at multiples of 4 is appropriate. She achieves this by inviting Andrew to consider a train with a unit of repeat consisting of 4 colours (by adding a green car). This move, which demonstrates an explicit choice of action to promote learning, can be seen as an example of her awareness-in-discipline. It is likely that the writer of this play saw major importance in this move, given that the character was named "Mrs. Green." In a play written by another prospective teacher, a similar intention was exemplified by drawing her student's attention to a pattern with a repeating unit RBYR, in order to distinguish from the given sequence RBY.

Excerpt 2: shift of attention toward language

T:	Why do you think the 80th car is red?
Zach:	Well, the red car is the 4th car, and we need to know what colour the 80th car will be so we divided 80 by 4 and it worked.
T:	What do you mean it worked? 80 divided by 4 = red car?
Zach:	Yeah, well no, like 4 divides into 80 evenly, so it must be red because the fourth car is red in the pattern.
T:	OK. So you are saying every fourth car in the pattern is going to be red, and 80 is a multiple of 4, so the 80th car must be red. Why don't you check your theory on some other multiples of four by writing out the pattern.
[...]	
Zach:	Red is wrong, a yellow car comes every three cars now, look! red blue yellow red blue yellow red blue yellow red 1 2 3 1 2 3 1 2 3 1
Hailey:	So the red car is actually the first car, we can't divide by 1
Zach:	So let's divide 80 by 3, what do we get?
Hailey:	Um, it equals 26.6666 Great, now what

Zach and Hailey's hands go up

Zack:	We decided that in the pattern every third car is yellow, so we divided 80 by 3, which is 26.6666. So we do not know what to do.
T:	You are on the right path, I'll give you a hint. Put away your calculators and divide it by yourself and think of it in terms or remainders...
Hailey:	Are you good at long division Zach?
[...]	
Peter:	Can I pretend the question asked me about 78th car?
Zack:	What for?
Peter:	if 3 goes into 78, 26 times evenly then the 78th car is yellow, and that is a multiple of 3, so to get to 80 you just add two more to the pattern, a red and a blue. So the 80th car is blue.
Hailey:	Oh my gosh, your right. You're so smart.

In the beginning of this excerpt we notice the writer's awareness of the language of mathematics. While Zach claims "like 4 divides into 80 evenly", the teacher picks up his idea, introducing appropriate mathematical expression "80 is a multiple of 4". Her mathematical-pedagogical suggestions are to "write out the pattern" and her hint is to "put away your calculator". In approaching the task, this nameless teacher's character has older or more sophisticated students than those of Mrs. Green. While in the previous excerpt the solution is found by modelling the train with manipulatives and then counting up by multiples of 3, in this play the solution is found by more efficient means: writing out the pattern and then carrying out the division. While the error that we introduced as the motivation for the play was observed with different age groups, including prospective teachers (Liljedahl, 2004), in the produced plays there are implicit assumptions about the students' prior knowledge and the approach is chosen to be appropriate for the students' level of mathematical sophistication. Further, in Haily's question, "Are you good at long division Zach?" the writer acknowledges her pedagogical awareness of the fact that many students consider long division as troublesome.

This writer also demonstrates her pedagogical awareness of different ideas students may bring to the task by introducing into the play some group work and collaboration among students. Further, the powerful idea of looking at a multiple of 3 close to 80 is voiced by a student. We note that not only does this idea lead to the desired solution, but also this student uses both the language accessible to his classmates – "3 goes into 30 26 times evenly" – and also the language modelled by his teacher – "that is a multiple of 3". We consider this delicate choice of wording as the writers' awareness-in-discipline. It appears intentional that the "smart" student's character is named Peter. ☺

Excerpt 3: shift of attention toward the learner

Novice teachers tend to go through a phase in which they focus on the curriculum to the exclusion of students' needs. Often this can be seen as a drive towards a specific learning outcome as opposed to working with the mathematics the student brings to a situation. In what follows we see a teacher who actually attempts to follow through with the students' mathematics.

Teacher:	Why is the 80th car red?
Stacy:	Because the 10th car is red. So, the 20th car, the 30th car, the 40th car, and so on will be red.
Teacher:	All right, so you are thinking that the patterns will repeat for each set of 10?
Stacy:	Yes.
Teacher:	Ok, well why don't you continue this pattern for a while and see another row until you reach 20 and then let me know if your theory is working out.
(5 minutes later)	
Stacy:	It didn't work ... the 20th car is blue!
Teacher:	Hmm ... okay. So are you still sure that the 80th car is going to be red?

Stacy:	Not any more. If the 20th car is different than the 10th car, then I'm not sure what the 80th car is.
Teacher:	So, what are you going to do?
Stacy:	I'm going to keep going to see what colour the 30th car is?
(1 minute later)	
Teacher:	Well …
Stacy:	The 30th car is yellow. So, I'm thinking that the 40th car will be red.
Teacher:	Why is that?
Stacy:	Because that is the pattern we have – red, blue, yellow.
Teacher:	Well, try it and see.
Stacy:	It is red. So that means the 50th will be blue, the 60th will be yellow, the 70th will be red, and the 80th will be blue.
Teacher:	Good … now why don't you go share your discovery with Aaron. I'm sure that he has something he wants to share with you as well.

We appreciate here the teacher's attempt to pursue the students' initial ideas in focusing on multiples of 10. However, while in the previous excerpts there was an attempt to have students understand the pattern, and not only to discover the pattern, in this excerpt the discovery about repeating pattern of colours within multiples of 10 remains unexplained. Recognizing that there are more efficient and transferable strategies the teacher mobilizes some of the knowledge that exists within the classroom by having Stacy work with Aaron – who, as it turns out, is using a more standard strategy of looking at the length of the repeating block. This is a valuable pedagogical shift of attention.

While in many examples the play writers exhibited a variety of shifts of attention, both mathematical and pedagogical, we still received a number of lesson plays where the lesson involved what we call "simple telling." In these imaginary lessons the teacher character simply corrected the student's error and explained how the problem should be approached. We suggest that the lesson plays prospective teachers write provide researchers with a window on mathematical and pedagogical knowing, and especially on their awareness-in-action, thus giving teacher educators a better opportunity to develop their own awareness-in-council. Our ongoing research analyses this mathematical and pedagogical knowing, thereby providing a novel contribution to the growing body of research on mathematics-for-teaching.

LESSON PLAY: TOWARDS 'REAL TEACHING'

"The key notions underlying real teaching are the structure of attention and the nature of awareness" (Mason, 1998, p. 244). However, much of the preparation for 'real teaching' is done in University/college classrooms. In attempting to make this preparation more effective, teacher educators continuously endeavour to design suitable tasks and experiences for prospective teachers. For example, Silver, Clark, Ghousseini, Charalambous, and Sealy (2007) advocate for 'practice-based professional development' and illustrate its key element, the 'professional learning task.' The goal of such a task is to engage prospective teachers in activities that resemble the daily work of practicing teaching. Professional learning tasks include examination of curriculum materials, video or narrative records of classroom teaching episodes and consideration of students' work. They "create opportunities for teachers to ponder pedagogical problems and their potential solutions through processes of reflection, knowledge sharing and knowledge building" (p. 262).

With the specific goal of developing teachers' reflective practice, Peng (2007) describes the task of 'lesson explaining' that has been designed in China. The task requires teachers "to explain how the content unfolds in the lesson, and the nature of mathematical challenges it offers" (p. 290). It includes explanation of mathematical content, justification of the chosen method and explanation of the 'teaching procedure', where the latter focuses on the development of students' ability, attending to learners' cognitive foundation and individual differences.

Biza, Nardi, and Zachariades (2007) developed tasks with "situation specific contexts," where they present prospective teachers with specific student response and seek their comments. The student response they chose was fictional, erroneous, and yet plausible. They suggest that such a task offers "an opportunity to explore and develop teacher sensitivity to student difficulties and needs as well as ability to provide adequate (pedagogically sensitive and mathematically precise) feedback to the student" (p. 303). They further acknowledge that since the engagement is not in the classroom and not in "real time" it provides teachers with the opportunity to think about their reaction and be reflective.

The task of lesson play fits well with the intentions and goals developed by the above-mentioned authors. It can be considered as a "professional learning task" (Silver et al., 2007) in creating resemblance to the work of practicing teachers. It includes components of 'lesson explaining' (Peng, 2007) and it starts with presenting fictional, yet plausible situations that are context specific (Biza et al., 2007) and seeks their resolution. Moreover, the task of designing a lesson play adds an important component in creating a situation of *"imagining* the real teaching", rather than simply discussing it. This is in accord with Watson and Mason's (2007) view that "the fundamental issue in working with teachers is to resonate with their experience so that they can *imagine* [our italics] themselves 'doing something' in their own situation" (p. 208). With this imagination, attention and awareness are developed in "slow motion," having a complete control of the situation and ability to replay or redress it, rather than "thinking on one's feet" and making in-the-moment decisions.

The task of creating a lesson play shifts prospective teachers' attention from general curricular objectives to specific teaching incidents, and invites them to imagine such incidents in a very detailed manner. Mason uses the phrase 'shift of attention' to focus on different mathematical aspects of a given problem. We adopted the phrase, but considered shifts that take place in teaching, which, in addition to shifts on various mathematical components of the task, include shifts to pedagogy, to didactics, to students' difficulties, and to language. Lesson plays provide an opportunity for prospective teachers to develop their awareness-in-discipline, to put in action their mathematical pedagogy, while attending to different demands of the complex act of teaching. We believe that creating these plays equips teachers with a repertoire of responses that they will be able to call upon in their 'real teaching'.

AFTERTHOUGHT

In a recent (March, 2008) meeting in Rome, celebrating the centennial of the International Commission on Mathematical Instruction (ICMI), Hyman Bass posed the following question to the participants of the working group on teacher education: "What does/should teacher educator know that an experienced teacher doesn't know?" It was not a simple question to answer and there was a feeling of unease in the group. After some discussion, Hyman Bass suggested that an experienced teacher should/does know how to design instructional tasks and the expertise of a teacher educator is in guiding teachers in their task design.

John Mason's opus, and specifically his work on awareness and shifts of attention, suggests a different answer. The difference between expert-teacher and expert-teacher-educator is not at the level of knowledge, but at the level of awareness. Awareness-in-discipline is a characteristic of an expert teacher and "what constitutes the practice of an expert" (1998, p. 260). Awareness-in-counsel is what guides the practice of teacher-educator. Designing tasks for students and guiding teachers in task design are just one example of these different levels of awareness in play. The lesson play task, in its various iterations, enriched our awareness-in-counsel while helping prospective teachers develop their awareness-in-discipline.

AFTER-AFTERTHOUGHT

In Hebrew "to pay attention" is *lasim lev*, a phrase that is translated literally as "to put a heart". This harmonizes nicely with Mason's (1998) suggestion that "we are our attention and we are where our attention is" (p. 251). It also gives enriched meaning to John Mason's suggestion that the significant products of mathematics education research are the "transformations in the being of the researchers" (Mason, 1998, p. 357). Indeed.

REFERENCES

Ashlock, R. B. (2005). *Error patterns in computation* (9th ed.). Upper Saddle River, NJ: Prentice Hall.
Biza, I., Nardi, E., & Zachariades, T. (2007). Using tasks to explore teacher knowledge in situation-specific contexts. *Journal of Mathematics Teacher Education, 10*(4–6), 301–309.
Liljedahl, P. (2004). Repeating pattern or number pattern: The distinction is blurred. *Focus on Learning Problems in Mathematics, 26*(3), 24–42.
Mason, J. (1998). Enabling teachers to be real teachers: necessary levels of awareness and structure of attention. *Journal of Mathematics Teacher Education, 1*(3), 243–267.
Mason, J. (1998). Researching from the inside in mathematics education. In A. Sierpinska & J. Kilpatrick (Eds.), *Mathematics education as a research domain: A search for identity* (pp. 357–377). Dordrecht, NL: Kluwer.

Mason, J. (2002). *Researching your own practice: The discipline of noticing.* London: Routledge.
Mason, J. (2007). *Mathematical pedagogy and pedagogical mathematics.* Presentation at the Pacific Institute for Mathematical Studies 10th anniversary, Simon Fraser University. Retrieved from http://mcs.open.ac.uk/jhm3/Presentations/%20Presentations%20Page.htm
Peng, A. (2007). Knowledge growth of mathematics teachers during professional activity based on the task of lesson explaining. *Journal of Mathematics Teacher Education, 10*(4–6), 289–299.
Pólya, G. (1945/1988). *How to solve it.* Princeton, NJ: Princeton University Press.
Silver, E. A., Clark, L. M., Ghousseini, H. L., Charalambous, C. Y., & Sealy, J. T. (2007). Where is mathematics? Examining teachers' mathematical learning opportunities in practice-based professional learning tasks. *Journal of Mathematics Teacher Education, 10*(4–6), 261–277.
Watson, A., & Mason, J. (2005). *Mathematics as a constructive activity: Learners generating examples.* Mahwah, NJ: Erlbaum.
Watson, A., & Mason, J. (2007). Taken-as-shared: A review of common assumptions about mathematical tasks in teacher education. *Journal of Mathematics Teacher Education, 10*(4–6), 205–215.
Zazkis, R., & Campbell. S. R. (1996). Divisibility and multiplicative structure of natural numbers: Preservice teachers' understanding. *Journal for Research in Mathematics Education, 27*(5), 540–563.
Zazkis, R., Liljedahl, P., & Sinclair, N. (2009). Lesson plays: Planning teaching vs. teaching planning. *For the Learning of Mathematics, 29*(1), 40–47.

Rina Zazkis
Simon Fraser University (Canada)

Nathalie Sinclair
Simon Fraser University (Canada)

Peter Liljedahl
Simon Fraser University (Canada)

PEDRO GÓMEZ AND MARÍA JOSÉ GONZÁLEZ

14. ANALYZING AND SELECTING TASKS FOR MATHEMATICS TEACHING

A Heuristic

INTRODUCTION

In this chapter, we propose partial answers to one of the main concerns in John Mason's opus: "Teachers set tasks because they believe that working on tasks will promote learning. But what learning might ensue from a given task, and how can tasks be selected to promote certain kinds of learning?" (Mason & Johnston-Wilder, 2004, p. 25). Mason's approach to this question is based on the ideas of behaviour, emotion and awareness, ideas that can enable the teacher to design and select tasks by helping him to organize questions to describe an individual's concept-image. We propose a complementary approach based on a heuristic procedure for analyzing the cognitive demands of school mathematics tasks and use this analysis to "psychologis[e] the subject matter" (p. 6). The purpose of the proposed procedure is to provide teachers with tools for analyzing and selecting tasks that promote students' development toward a learning goal. This type of analysis – that provides a method for conceptualizing the kinds of learning in a task – addresses one of Mason's main concerns regarding task interpretation and selection: "Teachers need to interpret tasks designed by other people and need to form views on what the intentions of the author were and what preparation the learners need if they are to benefit from undertaking the task. Teachers also need to be aware of how tasks might be interpreted by learners and know how to support the learners' activity in such a way that the experience can result in appropriate learning" (p. 8).

When a teacher undertakes the planning of a unit, he usually takes into account the curriculum design established for the topic at hand. This design might take different forms, but it generally defines a specific mathematical topic for which a set of learning goals has been established. A learning goal at this level is usually formulated as a statement like the following: "To recognize and use the graphical meaning of the parameters of the symbolic forms of the quadratic function and communicate and justify the results of its use." This statement can be considered one of several learning goals for the teaching and learning of the quadratic function topic in secondary school mathematics. It expresses the learning expectations that the teacher is supposed to have for his students concerning this aspect of the subject matter.

S. Lerman and B. Davis (eds.), *Mathematical Action & Structures of Noticing: Studies on John Mason's Contribution to Mathematics Education*, 179–188.
© 2009 Sense Publishers. All rights reserved.

The process of interpreting learning goals in terms of tasks and activities is not self-evident. Two different teachers can develop different interpretations of the same learning goal. Furthermore, a teacher's interpretation of a learning goal makes sense in practice only when he specifies its cognitive complexity through the choice of tasks that are aligned with a specific group of students in a particular context. Interpreting learning goals and setting up students' activities to achieve them thus requires knowledge and competencies that are not necessarily part of the teacher's knowledge base.

In this chapter, we propose a heuristic for analyzing and selecting tasks that can promote students' achievement of a given learning goal for a mathematical topic. "Since asking mathematical questions is vital, both as part of the presentation of mathematics and in the setting of tasks for students, it is appropriate to ask where these questions come from. Questions arise as pedagogic instruments both for engaging students in and assessing students' grasp of, ideas and techniques." (Mason, 2000, p. 97). Inspired by Simon's (1995) idea of a hypothetical learning trajectory, we seek to describe a procedure to enable teachers to characterize a learning goal in terms of its cognitive requirements and to analyze and select tasks based on that information. The procedure emphasizes the role in unit planning of the teacher's hypotheses about students' learning and their relation to the mathematical tasks he can propose in order to achieve a specific learning goal.

In what follows, we establish the meaning we give to the notions of learning goal and capacity. We then introduce the idea of learning path as a means for expressing the teacher's hypotheses about how tasks can promote his students' learning. We use these three notions for proposing a heuristic procedure that the teacher can use for analyzing and selecting tasks that can contribute to characterizing his students' development towards a learning goal. Finally, we discuss some of the contributions and constraints of such a proposed procedure.

LEARNING GOALS AND CAPACITIES

Learning goals express expectations of what the students should know and should be able to do as a result of instruction (Farrell, 1988, p. 196). These expectations can be formulated at different levels, from the national government's requirements to lesson planning. On the unit planning level, a learning goal is related to a particular mathematical topic in a specific course. Our interest centres on learning goals that involve "connected knowledge", as described by Mousley (2004), that is, the teacher's expectation that students will make connections between old and new knowledge, mathematical ideas and their representations, and academic concepts and real contexts. Although formulating learning goals is one of the most critical events in the instructional design process (Gagne, Briggs & Wager, 1994), we do not discuss the formulation of learning goals here. We will assume that the teacher has already chosen or has been assigned a specific learning goal. For instance, the Spanish education system gives the following as a learning goal for 16-year-olds studying the topic of surface areas of two and three-dimensional shapes:

Learning Goal (LG): To compute the area of figures given in real situations for which the data required in the formula are not directly known.

There is more than one way in which students can make progress toward a learning goal and construct the corresponding connected knowledge (Mousley, 2004). By selecting tasks, the teacher assigns the learning goal a specific meaning and establishes how he expects students to achieve it. He expects students to use their current knowledge and to establish new connections between that knowledge and the new ideas and techniques involved in the learning goal. Hence, the teacher manages two types of expectations: those expressed in the learning goal and those related to his students' current skills, which the teacher links to the students' concrete performance. We introduce the notion of capacity to capture the second of these expectations levels.

In the context of school mathematics we use the term capacity to indicate an individual's successful performance with respect to a specific assigned task (Gómez, 2007, pp. 64-66). We say that an individual has developed a capacity when he is able to solve the tasks requiring it. For instance, we say that an individual has achieved the capacity for developing plane surfaces when he successfully solves this particular task in a variety of standard situations (developing a prism, a pyramid, a cone). Capacities are specific to individual mathematical topics and are linked to types of tasks and observable student behaviours. The teacher decides whether a given piece of knowledge is a capacity based on his knowledge of the particular group of students to which it refers. Thus, what he considers to be a capacity for one group of students might be a learning goal for a different group of students.

Table 1 shows the 11 capacities identified for learning goal LG by a group of future teachers in an optional methods course implemented at the University of Cantabria during the academic year 2006-2007. The final project for the course was to design and justify a unit for the topic, "Surface areas of two and three-dimensional shapes". This group of future teachers followed the process we are

Id	Capacity
c_1	Recognizing the geometric elements (surfaces, segments, straight lines, etc.) to which a problem refers and drawing them.
c_2	Identifying the known and unknown data for a problem in 2D and 3D drawings.
c_3	Recognizing whether the Pythagorean theorem can be applied and knowing how to apply it.
c_4	Recognizing the properties of similar triangles and knowing how to apply them.
c_5	Knowing how to apply the properties of regular polygons.
c_6	Knowing how to apply the properties of triangles.
c_7	Knowing how to apply the properties of the circle and its circumference.
c_8	Knowing how to apply the formulas for area.
c_9	Transforming units of measurement.
c_{10}	Developing a surface in the plane.
c_{11}	Decomposing and/or completing a surface in order to compute its area.

Table 1. Capacities associated with learning goal LG

describing here to analyze and select tasks related to a learning goal. We will now use the work they did throughout the course to illustrate that process. The next section describes a procedure that the teacher can use for identifying the capacities associated with a learning goal.

In the context of planning, we believe that the notion of capacity plays the role of the basic procedural component for characterizing a learning goal. Developing the connected knowledge involved in a learning goal implies the appropriate coordination of a given set of capacities. These are the capacities the teacher associates with the learning goal, the basic skills that the teacher assumes his students already have and on the basis of which they can construct the new knowledge.

LEARNING PATHS OF A TASK

The teacher can select tasks based on his hypotheses about how those tasks can promote his students' learning. In most cases, the teacher can describe his hypotheses in terms of sequences of capacities. We introduce the idea of the *learning path of a task* as the sequence of capacities that students might put into practice when undertaking it. For instance, for the following task,

> t_1: Given a right triangle ABC with cathetus of length 8 and 6 units, respectively, we draw a parallel line to the minor cathetus, obtaining a new triangle APQ. The hypotenuse and the minor cathetus of triangle APQ are of length 5 and 3 units, respectively. Compute the area of the two triangles,

the teacher might expect his students to identify the geometric elements to which the problem refers, draw those elements, identify the known and unknown data, apply the formulae of area to ABC, apply the Pythagorean Theorem to APQ and apply the formulae of area to this triangle. That is, he might decide that a learning path of the task could be the following sequence of capacities (see Table 1):

$$c_1 \to c_2 \to c_8 \to c_3 \to c_8$$

A learning path of a task describes the sequence of capacities that the teacher expects his students to execute when facing the task. The teacher might expect his students to bring more than one learning path into play for a given task. For instance, if he considers that his students know how to relate the areas of similar figures, then he might decide that his students might also put into play the following learning path when solving task t_1:

$$c_1 \to c_2 \to c_8 \to c_4$$

The teacher can use the notion of learning path for analyzing and selecting tasks that can contribute to his students' development of the learning goal. In what follows we suggest a procedure that can give the teacher clues on how to do so.

ANALYZING AND SELECTING TASKS

We assume that a learning goal like LG has been assigned, for which the teacher wishes to select a sequence of tasks. In a first interpretation of the learning goal, the teacher can identify and delimit the mathematical content to which it refers. The teacher can produce a preliminary list of capacities related to the learning goal by performing a subject matter analysis of the topic at hand. Subject matter analysis is a procedure for analyzing a school mathematics topic in terms of the concepts and procedures involved in the topic, the multiple ways in which the topic can be represented, the phenomena that are related to the topic, and the relationships among these elements. The end result of the subject matter analysis can be expressed in concept maps that depict the variety of meanings that the topic acquires in school mathematics (Gómez, 2007, pp. 36–56). Subject matter analysis resembles "conceptual analysis" (Thompson, 2008) in the sense that it can be used in "building models of what students actually know at some specific time and what they comprehend in specific situations" (pp. 1–46). It thus provides information for identifying a list of capacities that might be associated with a given learning goal. However, these capacities must be linked to the teacher's expectations and hypotheses concerning the learning of a particular group of students undertaking a set of tasks.

Nowadays, teachers have access to a great variety of tasks for almost any topic (e.g., from experience, textbooks or the web). The issue is not to design new tasks, but to select or adapt those available to achieve the tasks most appropriate for the purpose at hand. The teacher thus starts the second phase of the process with initial sets of tasks T_i and capacities C_i. From T_i, he then chooses a task that he considers relevant and produces its learning path on the basis of the capacities in C_i. This learning path is a way of using his knowledge of his students to describe how he thinks that they would undertake the given task. In this process, the teacher might realize that the set C_i must be enlarged with new capacities or that some capacities in C_i must be formulated differently. For instance, when considering a task like the following,

t_2: *We want to line the lampshade of table lamp in the form of a truncated cone. The small and large circumferences measure 30 and 60 centimetres, respectively. The height of the lampshade measures 20 centimetres. How much tissue do we need?,*

the teacher might realize that his students ought to know how to perform developments in the plane or to identify situations in which the Pythagorean Theorem can be used. The teacher might have overlooked these capacities in his subject matter analysis of the topic. In this case, he can improve this analysis and enlarge the set C_i of capacities associated with the learning goal. Thus, task analysis in terms of learning paths can contribute to the teacher's identification of the capacities that characterize the learning goal. The teacher also analyzes a task based on the capacities he has identified. Analyzing the learning path of a task can lead the teacher to modify the task in order to include or eliminate a given capacity because he considers that it will not induce his students to bring into play the sequence of capacities he is interested in. For instance, when analyzing the task

t_3: *We want to tile the classroom floor. How many 30 × 40-floor tiles are needed to cover it?,* he might decide not to choose this task because he thinks that the context will allow his students to solve the problem by directly counting the floor tiles (without using any formula). The teacher can then perform a cyclic process of refining the list of capacities and selecting tasks. The cycle ends when the teacher decides that he has obtained an appropriate set of tasks T_m and capacities C_m: he expects the tasks in T_m to induce his students to establish links among capacities in C_m, thereby promoting development toward the learning goal.

The set of learning paths of a set of tasks T_m can be depicted on a graph. We illustrate the production of this graph with the work of the group of future teachers who proposed the list of capacities in Table 1 for the learning goal LG. They also proposed ten tasks related to these capacities and this learning goal. Table 2 shows two examples of the tasks they selected and their corresponding learning paths.

At this stage of their work, the group of future teachers produced the graph shown in Figure 1 for the learning paths of the ten tasks they selected. The graph shows the links between the capacities involved in those learning paths. The number of times that a link appears in the learning paths is indicated in parentheses. If there is no number, the corresponding link is brought into play in only one learning path. The capacities that represent the beginning of a learning path are indicated by a circle, those that represent the end of a learning path with a square. The number in square brackets indicates the number of learning paths for which this is the case.

Deciding whether the set T_m is appropriate for the learning goal involves a final step: a global analysis of the tasks in T_m. This analysis might produce information that cannot be obtained in the individual analysis of each task. For instance, when considering the complete set of learning paths that he expects his students to bring into play when solving the tasks in T_m, the teacher might realize that some links between capacities would seldom be used, whereas others would be executed frequently; that many tasks have the same starting or ending point; or that some capacities are never brought into play. This global analysis of the learning paths of the tasks in T_m is crucial. It shows the teacher how he is interpreting the learning goal in cognitive terms. It uses his students' cognitive performance to describe how he expects to promote development of the learning goal. It shows which cognitive

Example 1
A goat is grazing in a hexagonal area that is inscribed in a circumference with a radius of 10 meters. The goat is tethered to one vertex of the hexagon. How much land can the goat graze on if the rope with which it is tethered measures the radius of the circumference?
$c_1 \rightarrow c_5 \rightarrow c_2 \rightarrow c_{11} \rightarrow c_7 \rightarrow c_8$
Example 2
Compute the amount of tissue paper needed for constructing a kite formed of two sticks of length 75 and 50 centimetres, so that the short stick crosses the long one at 25 centimetres from one of its ends.
$c_1 \rightarrow c_2 \rightarrow c_8 \rightarrow c_9$

Table 2. Examples of tasks and their corresponding learning paths

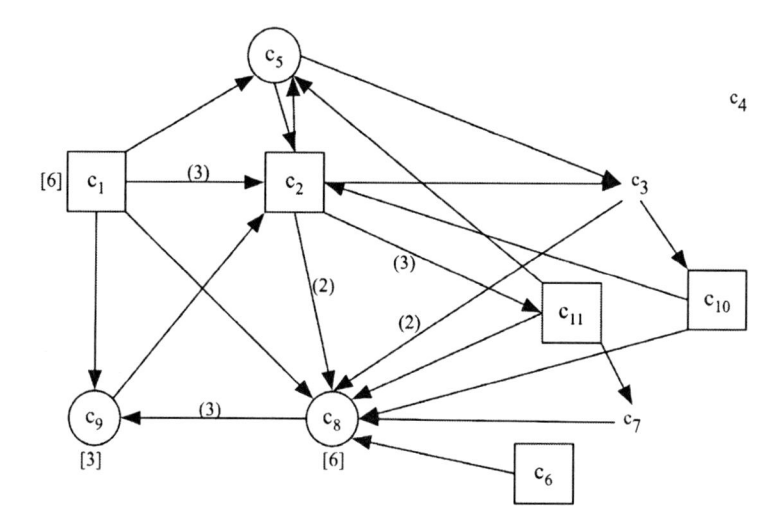

Figure 1. Graph of learning paths for the tasks selected related to LG.

requirements he is stressing in the tasks in T_m and which he is not. On the basis of this analysis, the teacher might decide to reconsider his choice of tasks in T_m and of capacities in C_m as a way of characterizing the learning goal. He might rule out some tasks in T_m and select new ones or even refine his choice of capacities in C_m. We illustrate this analysis using the graph in Figure 1.

The graph shows that capacity c_4 is not brought into play by any of the learning paths of the selected tasks, that most learning paths begin with capacity c_1 and end with capacity c_8, and that the links $c_1 \rightarrow c_2$, $c_2 \rightarrow c_{11}$, and $c_8 \rightarrow c_9$ are given greater emphasis. This type of analysis can provide the teacher with information about his interpretation of the learning goal. In the case of this example, it might show him the importance he gives to those three links. This information might induce him to modify the list of capacities he has chosen (e.g., eliminating capacity c_4 in the example above) and the set of tasks he has selected (reducing the emphasis on some links and increasing emphasis on others). The process of capacity identification and task selection ends when, at a given point in this cycle, the teacher considers that the resulting graph provides a good representation of his interpretation of the learning goal in terms of final sets of tasks (T_f) and capacities (C_f), that is, when he considers that the tasks selected are the best possible choice for promoting his students' learning. He can take this decision on the basis of the cognitive requirements of the tasks: the sequence of capacities that he expects his students to bring into play when undertaking the tasks.

LEARNING PATHS OF TASKS: CONTRIBUTIONS AND CONSTRAINTS

"In a sense, all teaching comes down to constructing tasks for students ..." (Mason, 2002, p. 105). We claim that the issue is not only to construct tasks but, from a set of tasks already available, to analyze and then select and adapt the tasks that, in the teacher's judgement, can best contribute to his students' attainment of a learning goal. Given that the teacher has a great variety of available tasks related to the learning goal, which sequence of tasks is the most appropriate to his context? The procedure we have presented provides information for making such a decision. This procedure is based on the assumption that the teacher knows the context in which instruction will take place. In particular, we assume that students' performance when solving tasks follows some patterns and that the teacher can predict them. Starting with the teacher's problem of selecting tasks that can promote his students' development of a learning goal, we have introduced the notion of learning path as a way of relating the teacher's predictions of his students' previous knowledge (the capacities), his expectations concerning the new knowledge they can construct (the learning goal), and the means with which he expects to promote such learning (the tasks). We believe that this notion and the heuristic we have suggested, enables the teacher to give a precise meaning (his meaning) to the learning goal at hand. The teacher expresses the meaning of the learning goal he assigns through the set of tasks that he believes will promote his students' attainment of it. This constitutes his prediction of how learning can evolve, based on his students' prior knowledge and the actions that he expects them to perform when solving the tasks.

In a constructivist setting, Simon and Tzur (2004) have suggested that students develop new knowledge as a consequence of iterations of activities linked to their effects. Elaborating Piaget's notion of reflective abstraction, they argue that, when students solve tasks, they create records of experience, sort and compare records, and identify patterns in those records. This mechanism provides a means for interpreting how students develop new concepts based on other, pre-existing ones. It is possible to interpret the links among capacities that are made explicit through the tasks' learning paths as the links that characterize the new connected knowledge expected in the learning goal. What is relevant when solving the tasks is the way the students might connect the capacities involved in order to arrive at a solution to the tasks and whether they become conscious of such connections. Students might develop the sequence of capacities brought into play when solving a set of tasks into a technique that can be used for solving other tasks, whether similar or not. They might do this, as Simon and Tzur suggest, by linking activities (performing sequences of capacities) and their effects (task solving). Achieving a learning goal means, among other things, becoming proficient in the recognition, selection and use of such techniques, so that what is a learning goal can now become, for another teacher and at a later time, a capacity that can contribute to the characterization of a different learning goal.

The proposed heuristic for task selection might offer a partial solution to the planning paradox (Ainley, Pratt & Hansen, 2006). According to the planning paradox, "if teachers plan from objectives, the tasks they set are likely to be unrewarding for the pupils and mathematically impoverished. Planning from tasks may increase pupils' engagement but their activity is likely to be unfocused and learning difficult to assess" (p. 23). By linking a learning goal to the tasks that can promote its attainment, the idea of the learning path can enable the teacher to see planning from objectives and planning from tasks as one coherent activity. We have assumed that most teachers are constrained to plan from objectives: they are expected to promote their students' attainment of learning goals that are set in advance. Yet this does not necessarily mean that the teacher should set unrewarding tasks. Based on his knowledge and prediction of his students' learning, the teacher can select mathematically powerful tasks that promote their development of connected knowledge. We have shown that the teacher can approach planning as a process that, for a given learning goal and a particular group of students, involves a cyclic revision of capacities and tasks. Furthermore, the idea of learning path and its corresponding heuristic provides the teacher with information about the cognitive links that can be fostered by the selected tasks. In practice, when interacting with his students, he can use this information for engaging them, adapting his decisions to their actions and assessing their performance.

Mathematics teachers' practice is far more complex than the description we have given in this chapter. We have simplified teachers' practice in order to focus attention on the specific steps that shape the proposed procedure. However, the origin and formulation of some learning goals might make them inappropriate for analysis with the tools proposed. Our concern here is with learning goals that are specific to a mathematical topic, precise enough to be worked on in a few lessons, and able to be formulated in terms of connected knowledge. The procedure assumes that the teacher has sufficient knowledge of his students to be able ultimately to determine the capacities that characterize the learning goal. Nevertheless, the teacher's assumptions might not be valid. He might find that his students have not developed one or more of the capacities he assumed they had. He will then need to change his planning in order to make sure that his students have the prior knowledge needed for reaching the learning goal. Further, planning involves other activities besides task selection as presented in this chapter. For instance, tasks can be classified in several ways (e.g., introduction tasks, motivation tasks, assessment tasks). That we have not taken those activities into account does not mean that we do not consider them relevant.

We do not expect practicing teachers to perform the learning path heuristic in detail. This activity requires more time than that usually available to the teacher. However, we believe that this procedure can be useful in teacher training. In such a setting, pre-service and in-service teachers can follow the heuristic and develop their competencies for performing it. Such an experience can then help the teacher in practice when planning lessons and units. It can enable him to interpret learning goals, give them a specific meaning adapted to his students, identify the cognitive requirements of the tasks available to him, and make explicit his expectations concerning his students' learning.

ACKNOWLEDGEMENT

This work was partially supported by Project SEJ2005-07364/EDUC of the Spanish Ministry of Science and Technology.

REFERENCES

Ainley, J., Pratt, D., & Hansen, A. (2006). Connecting engagement and focus in pedagogic task design. *British Educational Research Journal, 32*(1), 23–38.
Farrell, M. A. (1988). *Secondary mathematics instruction: An integrated approach.* Dedham, MA: Janson.
Gagne, R. M., Briggs, L. J., & Wager, W. W. (1994). *Principles of instructional design.* New York: Holt, Rinehart and Winston.
Gómez, P. (2007). *Desarrollo del conocimiento didáctico en un plan de formación inicial de profesores de matemáticas de secundaria.* Granada: Departamento de Didáctica de la Matemática, Universidad de Granada.
Mason, J. (2000). Asking mathematical questions mathematically. *International Journal of Mathematical Education in Science and Technology, 31*(1), 97–111.
Mason, J. (2002). *Mathematics teaching practice: A guide for university and college lecturers.* Chichester, UK: Horwood Publishing.
Mason, J., & Johnston-Wilder, S. (2004). *Designing and using mathematical tasks.* Milton Keynes, UK: Open University.
Mousley, J. (2004). An aspect of mathematical understanding: The notion of "connected knowing". In M. J. Hoines & A. B. Fuglestad (Eds.), *Proceedings of the 28th international conference of the international group for the psychology of mathematics education* (Vol. 3, pp. 377–384). Bergen, NO.
Simon, M. (1995). Reconstructing mathematics pedagogy from a constructivist perspective. *Journal for Research in Mathematics Education, 26*(2), 114–145.
Simon, M., & Tzur, R. (2004). Explicating the role of mathematical tasks in conceptual learning: an elaboration of the hypothetical learning trajectory. *Mathematical Thinking and Learning, 6*(2), 91–104.
Thompson, P. W. (2008). Conceptual analysis of mathematical ideas: Some spadework at the foundation of mathematics education. In O. Figueras, J. L. Cortina, S. Alatorre, T. Rojano, & A. Sepúlveda (Eds.), *Proceedings of the joint meeting of the International Group for the Psychology of Mathematics Education (IGPME 32) and North American Chapter (PME-NA XXX)* (Vol. 1, pp. 31–49). Morelia, MX.

Pedro Gómez
Universidad de Granada (Spain)

María José González
Universidad de Cantabria (Spain)

DAVID PIMM AND NATHALIE SINCLAIR

15. CULTURES OF GENERALITY AND THEIR ASSOCIATED PEDAGOGIES

INTRODUCTION

Twenty-five years ago, John Mason and David Pimm (1984) wrote an article "exploring meanings of 'generic' and 'generality'" (p. 277). In it, they offered and examined instances of these notions in everyday language use, as well as embedded in technical terminology that confronts the learner of mathematics. Their article ends with three sets of questions, all related to the pedagogy of teaching mathematics, none of which we would venture are any closer to receiving a convincing answer a quarter of a century later. Nonetheless, in this chapter we give these notions another look.

In his work, at various times, John Mason has attended closely to the mathematical, the historical and the pedagogical and especially how these three related elements intertwine. In this chapter, we wish to draw on instances from all of these arenas in order to start to examine the notion of a culture of generality and how, we wish somewhat grandly to claim, any text (in a general sense) from such a culture carries with it, more or less tacitly, an associated pedagogy. In mathematics, at different times and places, there have been arithmetic, geometric and algebraic cultures of generality, although nowadays students get rushed into algebraic modes very swiftly, as if they are perceived as the only legitimate ones.

By the term 'cultures of generality', we have in mind different forms in which mathematics has been presented historically (that is arithmetic, geometric and algebraic presentations rather than the cognate content areas of arithmetic, geometry and algebra – for instance, the predominant culture of ancient Greek mathematics is geometric even when the content is not geometry). But we also wish to consider more recent mathematical and educational manifestations, such as ones involving computer-based mathematics (dynamic), in addition to instances from cultures apparently outside mathematics (imagistic, poetic, aesthetic) – although the customary issues of space preclude anything but a fleeting glance at even one of these final varieties (the poetic). With poetry, one current aesthetic broadly eschews explicit generalities and abstractions, offering instead details, particulars, metonymically allowing the one both to stand for and to speak for the other. "This is that", asserts the poet, over and over and over.

Related to these ideas is the question of 'method' and what a method is and does. We have become increasingly interested in the origins of the notion of method and, in particular, what a method is within mathematics itself. In addition, there is a small cluster of related words, whose more careful delineation and distinction might lead to some clarity in wider realms, such as mathematics education. These words (in arguably decreasing order of specificity of application) are 'algorithm', 'method', 'technique', 'strategy', 'heuristic' and '*sutra*' (for a discussion of this final term, including the instance 'all from nine and the last from ten', see Joseph, 1992). All of the other words have Greek etymology, and 'strategy' has a nice inadvertent joke sitting inside it, as its (military) gloss is 'art of the general'. And it is worth recalling that 'art' has a significant sense of its own separate from 'the arts', just as 'skill' has a (sadly) disappearing sense separate from objectified and pluralised 'skills'.

ANCIENT PRACTICES

There are many places where arithmetic cultures can be seen historically at work: in India, in China, in the Middle East, and many of these texts of the past that we continue to have access to arguably have some pedagogic intent with regard to their readers. In consequence, they can be looked at from a pedagogic point of view, with questions of generality as well as technique or method in view. What is striking for us is that in the texts themselves issues of generality or generalization tend not to be specifically addressed, even though the specificities in certain cases are quite glaringly evident to contemporary eyes.

The Rhind mathematical papyrus, for instance, contains many instances of what to us nowadays seem to be 'worked examples', but what they are examples of is left completely tacit in the text (though that has not stopped historians and others whom one would expect better of to categorise them in quite anachronistic terms – e.g. the solution of first-order equations). Extensive and particular work with whole-number reciprocals as 'parts' makes us long for general results about decompositions and, arising from a different sort of longing, tempts us to attribute such general awarenesses or knowledge to them. A parallel with students in school following a sequence of such examples (without us falling into the trap of viewing ancient mathematicians as mathematically naïve) is not so far fetched.

Babylonian problem tablets again regularly contain several problems of a 'similar sort', where a method is perhaps in view, though unlike the Egyptian problems mentioned earlier, only the answer is given to each one (and very often is the same answer – below is one of 22 such problems, no working provided, but all involving a stone which whenever actually weighed was miraculously always 1 *ma-na*).

> I found a stone, (but) did not weigh it; (after) I subtracted one-seventh, added one-eleventh, (and) subtracted one-thir[tenth], I weighed (it): 1 ma-na What was the origin(al weight) of the stone? [The original(al weight)] of the stone was 1 ma-na, 9½ gin and 2½ se. (Fauvel and Gray, 1987, p. 26)

There is clearly much that could be said about this problem and its pedagogic presuppositions.

Here, now, are a few different examples from a single tablet, where an explicit procedure is broached. (The Babylonian numeration system is sexagesimal floating point – the ';' is a historical interpolation suggesting where the sexagesimal point 'should' be, had they used one.) Each problem details what an 'I' has already done and then provides what a 'you' is supposed to do in response. The presence in the text of personal pronouns allows a retention of a human voice and presence, someone (albeit possibly fictional) specifying the actual performance to be emulated rather than a theoretical action that simply *could* be done. We see no trace in these tablets of a general discourse of either processes or results.

1) I have added up the area and the side of my square: 0; 45. You write down 1, the coefficient. You break off half of 1. 0; 30 and 0; 30 you multiply: 0; 15. You add 0; 15 to 0; 45: 1. This is the square of 1. From 1 you subtract 0; 30, which you multiplied. 0; 30 is the side of the square.

2) I have subtracted the side of my square from the area: 14, 30. You write down 1, the coefficient. You break off half of 1. 0; 30 and 0, 30 you multiply. You add 0; 15 to 14, 30. Result 14, 30; 15. This is the square of 29; 30. You add 0; 30, which you multiplied, to 29; 30. Result 30, the side of the square.

3) (*actually number 7 on the tablet*) I have added up seven times the side of my square and eleven times the area: 6; 15. You write down 7 and 11. You multiply 6; 15 by 11: 1, 8; 45. You break off half of 7. 3; 30 and 3; 30 you multiply. 12; 15 you add to 1, 8; 45. Result 1, 21. This is the square of 9. You subtract 3; 30, which you multiplied, from 9. Result 5; 30. The reciprocal of 11 cannot be found. By what must I multiply 11 to obtain 5; 30? 0; 30, the side of the square is 0; 30. (Fauvel and Gray, 1987, p. 31)

But there are some nice pedagogic moments: for example, when there is a 'no-op' line (borrowing a computer term for a program statement that has no effect, other than to take time to execute, allowing the alignment of certain processes), the statement in the first question to the effect that 'One is the square of one'. In this particular problem, there is no point to such a line in the specific solution. But it acts as a placeholder for the fact that at this point *in the general solution* a square root is to be extracted; it gestures at the general. In addition, there is a nice ambiguity in the first problem too, when the 'I' imperatively instructs the 'you' to 'write down 1, the coefficient'. There are, of course, two salient coefficients, both of which are 1 here. A possible subtle pedagogical move may be at work here, suggesting that when there is a potential for ambiguity, deliberately invoke it in order to have students raise the question of 'which coefficient'. (For more details on these early problems, see Fauvel, 1987.)

Lastly, there is also a rhythm to these solutions, which is possible to emphasise poetically in terms of restructuring the solutions on the page (as a poem, perhaps – see Staats, 2008). Staats looks for similarity, parallels and repetition within a text. Here, treating each problem as a variation (on an undeclared, unstated, general theme) and then looking across 'stanzas' could produce a realisation of the sameness at work, the sameness that for us gets expressed in the completing the square algorithm (or 'algorhythm' perhaps).

Euclid

In Book IX of the *Elements*, Euclid presents his proof of the infinity of primes. The proof (by contradiction) has the familiar components and construction (although for an interesting restructuring of the 'logic' of a variation of this theorem, see Leron, 1983). Except that what it actually shows is that if there are three primes then there must be a fourth, different from the original three. There are no remarks about greater generality, no even passing hand-waving gestures of 'the other cases all go likewise'. We are simply offered a general claim and a specific proof. So in contrast to the Babylonian instance discussed above, the result is stated (twice) in its full generality (the *protasis* (enunciation) at the outset and again in the *sumperasma* (conclusion) at the very end of a theorem – see Netz, 1999, pp. 10–11). The actual proof (*apodeixis*), however, is offered in its full specificity.

Richard Courant, talking about David Hilbert's consciously used principle for working on mathematical problems, observed:

> If you want to solve a problem, first strip the problem of everything that is not essential. Simplify it, specialize it as much as you can without sacrificing its core. Thus it becomes simple, as simple as it can be made, without losing any of its punch, and then you solve it. The generalization is a triviality which you don't have to pay much attention to. (1981, p. 161)

It seems to us that what Euclid has done might be seen in such a light, but we are nevertheless still struck that nothing whatsoever is said in the text. We turn now to a fuller consideration of a similar phenomenon in a possibly less well-known example.

Archimedes

The first proposition of Archimedes' *On the Sphere and the Cylinder I* claims that if a polygon circumscribes a circle, the perimeter of the circumscribed polygon is greater than the perimeter of the circle. The statement appears very general, applying to any given polygon, and is stated in terms of 'polygon'. In the written text of the proof, no mention is ever made of a particular polygon. However, the diagram included with the proof shows a circle circumscribed by a pentagon. Looking carefully at the proof, which involves showing that at each angle of the pentagon the length of the two subtending chords is greater than the length of the intervening arc on the circle, we see that Archimedes mentions each of the five angles before concluding that "therefore the whole perimeter of the polygon is greater than the circumference of the circle". Even though his proof generalizes beyond the pentagon (since one merely has to take into account each angle of the polygon in turn), Archimedes makes no explicit statement about the link between his particular analysis, with its connection to the particular diagram, and the more general proposition. The proof speaks an apparent generality while pointing at a particularity. In leaving this confusion between word and referent, there is a tacit invitation to notice this confusion and thereby ponder the reasons for it.

Despite the linguistic gesture toward the general, Archimedes remains in the specific both in the actual analysis of the proof, and, especially, in the diagram itself. The specificity of the diagram is not surprising; if one wants to create a diagram, one must choose a specific polygon. No diagrammatic equivalent of the word 'polygon' is possible, when you are actually working mathematically with the diagram, as the ancient Greeks did. Netz (2004) argues that in choosing the pentagon, however, Archimedes was practicing a complex form of generality common among ancient Greek geometers. Archimedes could have chosen a triangle, which would have been simpler, or even a square. But simple shapes may be dangerous, carrying with them 'special' properties that would not generalize to more complicated shapes. In this sense, the pentagon is the simplest of non-special shapes – a generic case perhaps – and thus an obvious candidate to represent some form of generality.

Netz also points out that Archimedes could have made the generality of the proof explicit in several ways. For example, in listing the different angles of the polygon, he could have chosen a specific order (clockwise or counter-clockwise) in order to suggest a repeatable strategy applicable to any polygon. Instead, Archimedes lists the angles in a seemingly erratic way. Alternatively, he could have explicitly stated that comparison of angles to arcs could apply to polygons with greater (and lesser) numbers of sides. Since he does not, we are left to believe that, at least in part, the diagram – and the careful choice of the pentagon over any other polygon – carries the weight of the responsibility for *showing* the generality of the statement. Diagrams would have thus played quite a different role in the culture of the ancient Greeks than they do in modern mathematics. As Netz argues, diagrams were schematic, used to show the logical structure of a geometrical configuration rather than a materially correct picture. As a schematic representation, Archimedes' diagram of the pentagon achieves the generality that more linguistic approaches would have done, and in a rather more direct way.

Diophantus

An etymology for the Greek word *methodos* as *meta+hodos* seems plausible, where *hodos* (οδος) can refer prosaically to 'street' or 'road', but also to 'way'.[1] This is a place where research in the history of mathematics can prove more broadly informative. For example, Christianidis (2007) details the results of his careful linguistic-mathematical study of Diophantus' *Arithmetica* from late Antiquity, paying close attention to Diophantus' assertion that to solve arithmetical problems one should "follow the way [*odos*] I will show" (p. 289). Christianidis sets out to examine a persistent issue in the literature on Diophantus:

> the question of whether or not Diophantus elaborated and employed a single general strategy for the treatment of arithmetical problems. To my mind, this issue cannot be adequately treated without prior clarification of another issue: the exact description and characterization of the mathematical practice of Diophantus. (p. 291)

In addition, Christianidis seeks to weigh in on a much-discussed boundary dispute about whether and when this work of Diophantus is algebra or arithmetic, as well as offering a cogent view on why Diophantus frequently posed a problem type in general and then only proceeded to solve a single particular instance of each type (thereby echoing both Euclid's strategy described in the above discussion and the Babylonian approach to communicating a method).

> II.8 To divide a proposed *tetragônos* [square number] into two *tetragônoi*.
>
> Let it be proposed to divide 16 into two *tetragônoi*. [...] The one will therefore be 256/25, the other 144/25 and their sum is 400/25, and each is *tetragônos*. (p. 300)

And, as this article makes clear, to speak of Diophantus' *method* is to do a considerable disservice to the subtlety of precisely what this mathematician was and was not offering his readers.

Problems of particular and general beset discussions of method, not least as, for some, generalisation is presumed to be the only game in town (as compared with appropriability, for example, or even enlightenment). Diophantus' solutions can be profitably seen in this light. And as this brief taste might indicate, Diophantus' manner of public working has something to offer even an initial exploration of cultures of generality.

MODERN PRACTICES

As mentioned previously, the ancient Greek diagram was used as a schematic to represent the logical structure of geometric configurations. We see the recent emergence of dynamic geometry software as, in some ways, harking back to this ancient culture of generality. In dynamic geometry environments, such as *The Geometer's Sketchpad* or *Cabri-géomètre*, materially correct *drawings*, such as the one shown below (Figure 1, left), for an equilateral triangle, are contrasted with *diagrams*, which encode the mathematical relationships of the geometric object, and which, unlike the drawing, remain equilateral triangles under dragging. As shown below (Figure 1, right), the drawing of an equilateral triangle breaks down when dragged, retaining only the relationships that define a triangle.

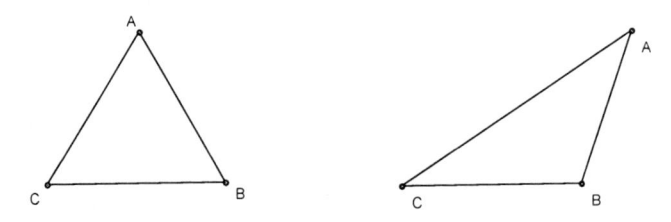

Figure 1. A drawn equilateral triangle (left); the 'same' triangle after dragging A horizontally to the right (right).

CULTURES OF GENERALITY

While static representations exist for both drawings and diagrams, their dynamic counterparts have both a pragmatic and a mathematical benefit. Pragmatically, one can empirically test whether the diagram is correct by dragging it. Mathematically, the specific diagram (shown in Figure 2) gestures toward generality by virtue of the fact that, materially, it is any, or perhaps every, equilateral triangle.

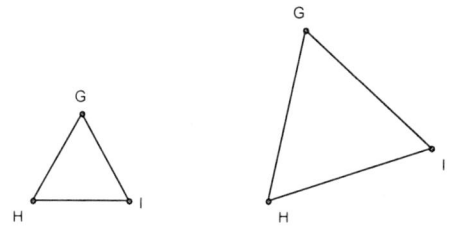

Figure 2. Equilateral triangle (left); the 'same' triangle after dragging I (right).

Doing mathematics with dynamic geometry shifts attention from *what is* the object or the relationship, to *what happens* to the objects or relationship under continuous dragging. Consider the diagram of the orthocentre of a triangle. The definition of the orthocentre is encoded in its construction – the point where the three altitudes meet. By dragging one of the vertices of the triangle (see Figure 3), the user prompts the question of what will happen: as the vertex moves down the screen, making the triangle more obtuse, the orthocentre exits from the triangle.

Indeed, one might conceive of, as Mason (2005) does with different image frame sets, asking students to figure out what the film might look like that relates the two diagrams in Figure 3. Such a request might draw students' attention to the fact that the orthocentre needs somehow to fall on the triangle in its journey from inside to outside. In using *Sketchpad* to generate the 'film', the continuity of the dragging, and the changing geometric configuration, reveal the doorway as the interesting crossover point between acute and obtuse triangles – where the triangle is right-angled and where O falls exactly on vertex A. It seems worth considering whether the ability to generate the kind of film Mason describes depends on having already been able to imagine a triangle in motion, an ability that is not supported by the static media of school mathematics – and a mathematical move that was highly suspect for many mathematicians, ancient and modern.

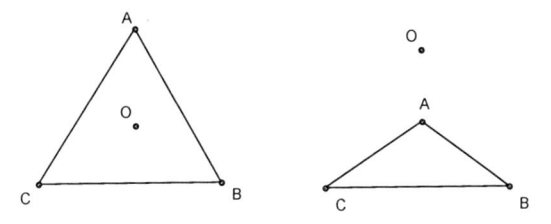

Figure 3. The orthocentre (O) of triangle ABC (left); O after dragging A down (right).

In the culture of dynamic representations, generality seems to depend squarely on time, and on how things change continuously in time, under dragging. This contrasts both with the visual generality of the ancient Greeks and the linguistic one of modern mathematics. Diophantus could continue to generate other particular solutions to various of his problem types (e.g. in II.8 as given above, he could work with finding other (particular) solutions for starting *tetragônoi* other than 16) for as long as he wants. The potential of continuing in time promises a greater generality.

Pedagogically, dynamic geometry representations change the traditional approach learners take through mathematical conceptualization. Take the triangle, for example. If previous approaches, based on the availability of a limited number of static drawings (usually with the base of the triangle parallel to the bottom of the sheet of paper, and drawn within the constrains of the printed page), encouraged the learner to develop a richer example space containing more, diverse examples of triangles, as Watson and Mason (2005) argue, then this newer approach goes in the opposite direction. With the draggable triangle, the learner can effortlessly generate many, diverse triangles, and the focus turns to which ones are specifically of interest, or perhaps, which ones can be seen as generic examples. As Pimm (1995) notes, the move from one particular to another is so easy that the particular status of a diagram may in fact be emphasised over the general one: dynamic diagrams may value seeing "the one as one *among* many" instead of "seeing the one *as* the many" (p. 59; italics in original). The shift is characteristically post-modern, going from the general to the particular, from the property to the behaviour, from the normative to the idiosyncratic.

In dynamic geometry representations, the point (and all objects built on it, like the polygon) must initially be chosen to have a particular location, but that location is always amenable to change. The specific point is pre-loaded with potential generality. Similarly, the choice of the parameter a in $ax^2 + bx + c$, is committed to generality, as is the variable x. Mason and Pimm (1984) draw attention to the difficulty of distinguishing the parameter from the variable in such an expression, and, in particular, of navigating through the generals and specifics of a question such as "hold x fixed, e.g. put $x = x_0$, and let a vary – what happens …?" (p. 284). By contrast, *Sketchpad's* dynamic parameter demands an initial specific choice (say $a = 2$), but, once again, it pre-loads that specific choice with generality in that the 2 can be *any* other number. The expression is now one that has potential. And the epistemic difference between the variable and the parameter is not just one of perception, but also one of time-dependent, continuous action and agency.

The symbolic representations offered by algebra have a powerful way of sweeping away particularities, as can be seen in the quadratic expression given above. Thus, in the formula for the area of a triangle, $A = bh/2$, the variables b and h not only stand for attributes of any triangle, but also for any of the three combinations of bases and heights that a given triangle might have. (For more on this and its connection with school algebra, see Proulx & Pimm, 2008.) The simple formula obscures the double generality – and, especially for students, the latter one. By interpreting the formula geometrically, we can shift our attention away from this latter generality.

CULTURES OF GENERALITY

Consider the three rectangles formed by the three pairs of bases and heights of the given triangle ABC. Visual comparison of the three rectangles produced by each *bh* pairing should leave some doubt as to whether their areas are indeed the same (see fig. 4). The sameness of the formula contrasts with the distinctiveness of the geometry. Dragging ABC leads to the realisation that the rectangles are really only visually the same under special circumstances. The algebraic mode may be more persuasively general; however, the experience of seeing three non-congruent rectangles continue to have the same numerical area can be a pleasant surprise, and can lead one to appreciate the insights offered by, instead of going from the particular to the general, sometimes travelling in the opposite direction – a theme we will discuss in more detail.

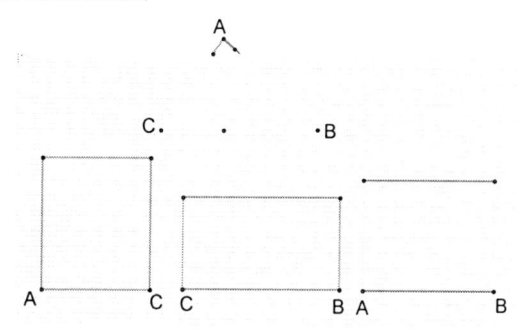

Figure 4. The three different ways of representing the double-area bh of the triangle ABC.

We have considered dynamic geometry as reflecting a culture of generality that depends very much on its technology-based representational system. Unlike the cultures mentioned above, which defined the mathematics of their time, dynamic geometry evolved from and continues to be motivated by pedagogical concerns. While many mathematicians use dynamic geometry in their own research, the current culture of mathematics continues to be dominated by the linguistically and symbolically driven algebraic culture of generality. As Rotman (2008) has powerfully argued in his recent discussion of technologised mathematics, with the disappearance of the alphabet and its replacement by virtual, visual, parallel-processed images (thereby shedding the sequentiality and linearity that characterise the alphabet), the *mathematician* will change and this will certainly change the culture of generality in which we currently work mathematically.

Despite their differences, the dynamic geometry culture also shares common features with arithmetic, geometric and algebraic cultures of generality. We have already discussed the special role of the diagram in the move toward generalization of the geometric culture of the ancient Greeks. We note that while the ancient Greeks worked without algebra, the computer screen on which dynamic images move are driven by algebraic calculations and transformations. Since these are hidden, the visible images enable the learner to work on mathematics through time rather than text, achieving Dick Tahta's pseudo-Daoist criterion for what it means to be doing geometry: 'the geometry that can be told is not geometry'.

As with arithmetic cultures of generality, we see the similar compulsion to do mathematics with a particular example (not necessarily a generic one). Moreover, the 'method' for solving, say, a problem involving decomposition, might be compared with the 'method' used to construct an equilateral triangle in *Sketchpad*. Each method produces one object (one solution, one particular triangle that depends on two particular points), even though it might be understood that any such object could be produced. Of course, the solution to the arithmetic problem marks the end-point of inquiry, so that method is one thing potentially amenable to generalisation, whereas the constructed triangle is just the beginning of inquiry.

POETIC PRACTICES

> What is a poem, after all, if not the consequence of ruminations on the interplay of watcher and watched ... a poem is definition, theorem and proof all in one. (Joan Geramita, a mathematician at Queen's University, in Page, 2003, p. 58)

In an interview with the poet Donald Hall, published in 1959 in *The Paris Review*, T. S. Eliot was asked for 'advice to a young poet about what disciplines or attitudes he might cultivate to improve his art'. He replied:

> I think it's awfully dangerous to give general advice. I think the best one can do for a young poet is to criticize in detail a particular poem of his. Argue it with him if necessary; give him your opinion, and if there are any generalizations to be made, let him do them himself. (p. 19)

This quotation of Eliot's rings with a comparable injunction of Caleb Gattegno to the effect that the teacher should never make a remark that a student is capable of making. But it also evokes the idea of working powerfully through the specific.

There seem at least two pertinent points of reference with poetry, one with regard to the aesthetic of eschewing as far as possible stating the general (in general) – though it is very much in the mind and eye of the poet – and the other with regard to teaching using model poems. With regard to having to 'earn' a generality in a poem, this goes hand-in-hand with an often-stated injunction to 'show don't tell'. This latter practice, one that could perhaps usefully be emulated by mathematics teachers at all levels, leads to the specificity of images, of settings, of the consummately chosen 'telling' detail.

In our earlier discussion of Archimedes, we mentioned the specific 'show' aspect of the diagram and the 'tell' aspect of the accompanying language, and the challenge of precisely calibrating the two. With *Sketchpad*, it is possible for the computer to 'do the generalising' in that as a user carries out a specific construction sequence of commands, the program generates the generalized text that enables a general procedure to be created, irrespective of the state or degree of awareness of the user.

When working from model poems, the modelling might be of various aspects: the structure, the format, the grammar, the diction, the content, the embedding of a question, ... – simply some feature of the poem to be taken as a guide, as a parametric possibility. But as we mentioned in the opening, even to *notice* or *see* a feature implies some generalisation has already been made. To speak of attribute of an object is to have generalised it. See Appendix 1 for two instances of poems staying very close to the form of the model poem. Despite one of us having written them, he is unable to say whether he went through a general in order to create a 'likeness' – working, in Mulcaster's lovely phrase, with "the likeness of unlike things". They are offered here in the light of John Mason's fondness for reader tasks, with the invitation to look for the particular sense of simile at work. But this brings into sharp relief the question of what someone takes from looking at a specific instance, and how a teacher might help a student attend to features she or he would like them to be able to discern.

Seamus Heaney claims:

> Every writer lives between the vernacular given – whether it be the vernacular of Oxford or of the Caribbean – and some received idiom from the tradition. Ted Hughes had a marvelous little parable about this. Imagine, he said, a flock of gazelles grazing. One gazelle flicks its tail and all the gazelles flick their tails as if to say, "We are eternal gazelle". (in O'Driscoll, 2008, p. 447)

> [...] if a poem is any good, you can repeat it to yourself as if it were written by somebody else. The completedness frees you from it and it from you. You can read and reread it without feeling self-indulgent: whatever it was in you that started the writing has got beyond you. (in O'Driscoll, 2008, p. 197)

In this observation, we see an authorial detachment from the product of their work, a stepping away from who wrote the poem and when (depersonalization and detemporalisation, two of Nicolas Balacheff's three characteristic requirements of mathematical proof writing), both stages of detachment on its tentative, troubled journey *en route* to the eternal.

CONCLUSION

'The concrete is the abstract made familiar by time.'

This observation is often attributed to Hadamard (though we have been unable to track down the source), and it draws attention to how hard it can be, at times, to see the particular, when you do not have a sense of the general of which it is an instance. Writing about cultures of generality here has raised in a number of places the connection to time: historic time, narrative time, mathematical time. Time has had a very complex relation to mathematics, not least in terms of the dynamic and any manifestation of motion in proof.

Philosopher-poet Jan Zwicky (2003), in her elegant and lyrical work on metaphor, much of it in relation to aspects of mathematics, makes this additional connection, embedding it in a link between generality and beauty:

To realize it could be any right triangle, any square, is to experience the beauty of a mathematical truth. To grasp a geometrical truth is to grasp a gesture that is meaningful in an enormous array of contexts – in fact, all that are available to the spatial imagination.

The experience of beauty is the experience of some form (or other) of relief from time. (2003, p. 71 left)

Zwicky has also written on lyric as contrasted with narrative poetry, characterizing the former in terms of a different 'syntactic glue' from the latter. Lyric's connectivity is one of accretion of particulars: this AND this AND this AND ..., while narrative is connected sequentially: this THEN this THEN this THEN ... (see Zwicky, 2006). For us, mathematics could be captured as: this THERFORE this THEREFORE this THEREFORE ... and, as such, might seem to be the furthest away from a lyric sensibility. However, in thinking back to the juxtaposing, the concatenating of particulars, of Babylonian examples placed side-by-side, of Sketchpad figures morphing one into another, we can see how a theorem is at one and the same time asserting this case AND this case AND this case AND ..., and in so doing is attempting to give us, in Zwicky's beautiful phrase, "the experience of some form (or other) of relief from time".

In this spirit, we end with a poem.

LYRIC MATHEMATICS
Lyric attempts to listen [...] without imposing
a logical or temporal order on experience.
This, it says, *This, and this and this.* (*Jan* Zwicky)

Beyond the wake of a vision,
a theorem is a calm,
an equanimity,

the circle of still water
that follows a humpback's fluke,
a lee from this rippled world.

A theorem is a lyric that lists as it listens,
every case that is the case, all at once.

A theorem is a likeness of unlike things,
both list and a listening to indifference.

APPENDIX 1: AT PURI, THE CROWS

The first poem here, by Jayanta Mahapatra (1972, p. 341), is the original, used as a model. The two that follow are based on it.

TASTE FOR TOMORROW

At Puri, the crows.

The one wide street
lolls out like a giant tongue.

Five faceless lepers move aside
as a priest passes by.

And at the street's end
the crowds thronging the temple door:
a huge holy flower
swaying in the wind of greater reasons.

AS IF THERE WERE NO TOMORROW

At Juan-les-Pins, the jazz.

The grove of park pines
silts down to the shore.

A sextet of cicadas rub awake
as a whitebeard bluesman eddies by.

And at the saltwater's edge
the reverent ring the stage:
a brazen peacock's tail
fanning the heroic flames of hell.

THE FLOOD

In my kidney, the stones.

The high cave roof
narrows like some wicked chicane.

Five small stalactites snap off
as the boulder spelunks by.

And at the portal's end
the mass bleeds the temple walls:
huge holy stigmata
cleansing in the Ganges of past excess.

NOTE

[1] When John (14:6) reports Jesus saying, "I am the way, the truth and the life", 'way' is οδός in his original Greek. And οδός, like *via* in Latin and unlike *dao* in Mandarin, is attached to a human being.

REFERENCES

Christianidis, J. (2007). The way of Diophantus: Some clarifications on Diophantus' method of solution. *Historia Mathematica, 34*(3), 289–305.
Courant, R. (1981). Reminiscences from Hilbert's Göttingen. *The Mathematical Intelligencer, 3*(4), 154–164.
Eliot, T. (1959). T.S. Eliot: The art of poetry, no. 1. *The Paris Review, 21*, 1–25.
Fauvel, J. (1987). Unit 1, *MA290 Topics in the history of mathematics*. Milton Keynes, UK: The Open University.
Fauvel, J., & Gray, J. (1987). *History of mathematics: A reader*. Basingstoke, UK: Macmillan Education.
Joseph, G. (1992). *The crest of the peacock: Non-European roots of mathematics*. London: Penguin.
Leron, U. (1983). Structuring mathematical proofs. *The American Mathematical Monthly, 90*(3), 174–184.
Mahapatra, J. (1972). Taste for tomorrow. *The Critical Quarterly, 14*(4), 341.
Mason, J., & Pimm, D. (1984). Generic examples: Seeing the general in the particular. *Educational Studies in Mathematics, 15*(3), 277–289.
Mason, J. (2005). Mediating mathematical thinking with e-screens. In S. Johnston-Wilder & D. Pimm (Eds.), *Teaching secondary mathematics with ICT* (pp. 219–324). Maidenhead, UK: Open University Press.
Netz, R. (1999). *The shaping of deduction in Greek mathematics: A study in cognitive history*. Cambridge, UK: Cambridge University Press.
Netz, R. (2004). *The works of Archimedes* (Vol. 1). Cambridge, UK: Cambridge University Press.
O'Driscoll, D. (2008). *Stepping stones: Interviews with Seamus Heaney*. London: Faber and Faber.
Page, J. (2003). *Persuasion for a mathematician*. Toronto, ON: Pedlar Press.
Pimm, D. (1995). *Symbols and meaning in school mathematics*. London: Routledge.
Proulx, J., & Pimm, D. (2008). Algebraic formulas, geometric awareness and Cavalieri's principle. *For the Learning of Mathematics, 28*(2), 17–24.
Rotman, B. (2008). *Becoming beside ourselves: The alphabet, ghosts and distributed human being*. Durham, NC: Duke University Press.
Staats, S. (2008). Poetic lines in mathematical discourse: A method from linguistic anthropology. *For the Learning of Mathematics, 28*(2), 26–32.
Watson, A., & Mason, J. (2005). *Mathematics as a constructive activity: Learners generating examples*. Mahwah, NJ: Erlbaum.
Zwicky, J. (2003). *Wisdom and metaphor*. Kentville, NS: Gaspereau Press.
Zwicky, J. (2005). Lyric, narrative, memory. In R. Finley, P. Friesen, A. Hunter, A. Simpson, & J. Zwicky (Eds.), *A ragged pen: Essays on poetry and memory* (pp. 87–105). Kentville, NS: Gaspereau Press.

David Pimm
University of Alberta (Canada)

Nathalie Sinclair
Simon Fraser University (Canada)

ANDY BEGG

16. NOTICING

Looking Forwards and Backwards

INTRODUCTION

England is a long way away from New Zealand where I have lived for most of my life, but for over twenty years John Mason influenced my work and the work of colleagues in New Zealand and Australia in mathematics education and more generally. Over these years we met periodically at international conferences, and recently I worked with him for two years at the Centre for Mathematics Education at the Open University on a project for retraining teachers that had a focus on mathematical thinking. At the same time I completed a doctoral project (Begg, 2008) that he and Linda Haggarty, another friend and colleague, supervised. Over these years I have been influenced by the three complementary strands of John's work—thinking, noticing, and variation. In this chapter my intention is to discuss some of my work and the influence that *noticing* has had on it.

Fairly early when discussing the *discipline of noticing* I had the impression that one noticed what was happening to prepare and anticipate for the future, I remember John saying something like:

> I was watching a teacher and noted an incident. I wondered what I would do if this happened to me, I noted the incident in my journal, thought about it, and noted how I would handle the situation. About 15 months later it did happen, I knew what to do, and it worked.

While working with John and Linda I had decided to use autobiography in my research and it was soon evident that I was using *noticing* in a reflective rather than an anticipatory way—remembering incidents from the past, noticing how these seemed to signify changes in my ways of working, and thinking how these changes had impacted. Of course reflective noticing is not significantly different from anticipatory noticing; both involve thinking about an incident in a way that might be useful in the future. Part of such reflective noticing involved thinking about some of the main influences on my work.

INFLUENCES

The first liberating book that I read about education was *Summerhill* (Neill, 1962). This changed my stance in the classroom to one which involved a more equitable relationship between my students and me. It gave me confidence when I was working in an alternative school, and led me to work in more anarchic ways as a curriculum officer when I had significant responsibility for high school mathematics in New Zealand.

The next two major influences on my thinking were the *Cockcroft report* (1982) and the *Standards* (National Council of Teachers of Mathematics, 1989). The *Cockcroft report* made me seriously question what we were doing in schools; and the *Standards* caused me to formally consider mathematical processes alongside content topics, that is, the balance between doing and knowing.

These two influences also suggested the need to move from behaviourism to constructivism, and at the time some of my colleagues (in both mathematics and science education) had been discussing various aspects of the theory. I tended towards radical constructivism (von Glasersfeld, 1991) while my science colleagues preferred social constructivism, though I thought that although individuals construct their unique meaning, this was within a world-view that was influenced by culture and society and that enough was constructed in common to enable satisfactory communication. Later my views about learning shifted further as I read about bodily knowing, enactivism, emergence, and complexity (Maturana & Varela, 1987; Varela, Thompson & Rosch, 1991; Davis, 1996; Varela, 1999).

My areas of interest had moved from curriculum and professional development to the nature of mathematics, learning theories, ethno-mathematics, and then back to curriculum, and, while working at the Open University, to thinking.

Noticing

The first thing I noticed on arriving at the Open University and entering John's office was that his collection of books included most of those that were not 'mainstream' but had been significant for me, as well as many others. I anticipated many interesting discussions and was not disappointed.

While listening to John in the past and working with him at the Open University I noticed his very practical focus on what we teach and how we teach. In my work with teachers I had spent considerable time considering and discussing underpinning theories and attempting to justify my choices from theoretical perspectives. It quickly became obvious to me that John's theoretical background was very much richer than mine and seemed more integrated, yet his focus of teaching through examples (tasks or activities) was much more appealing to teachers than my approach had been.

The retraining project we were working on was intended for teachers who had either taught mathematics many years earlier or who had taught subjects other than mathematics. We realized that these teachers did not want a refresher course on school or undergraduate mathematics, and that we needed to take a different approach to engage them with the subject in a way that they might engage their pupils in the future. The focus that had been chosen was mathematical thinking, but rather than talk about thinking we wanted to stimulate it.

One of my favourite tasks that John suggested as suitable for inclusion was:

Can you cut a square into exactly 11 squares?
Explore.
And then some possible follow-up questions:
If not, can you prove it is impossible?
If yes, can you do it in another way, and another, and …?
Then finally:
Generalise.

I found this task fascinating. It is one of the very few mathematical tasks that I have worked on for well over 40 hours. Typically 13- and 14-year-olds find an initial solution in about a minute, while mathematics teachers usually take 90 seconds and other teachers a little more. Then most stop, which seems to imply that for them mathematics has always been finding the solution.

As I worked on this task I found that the possibilities were wonderful. I have 16 solutions so far, but there may well be more. I noticed a number of other prompt questions emerging, but when using this task with teachers and students I have always been concerned whether asking these may simply be me privileging my way of thinking. These prompts include:
- What makes solutions equivalent (congruence/similarity/'jigsaw' equivalence.)?
- Can it be done with squares of 1 size, 2 different sizes, 3, 4, … ?
- Is it useful to approach the task numerically rather than visually? And for generalizing
- What numbers other than 11 is it possible for? (Convince someone else.)
- What shapes other than squares does it work for?

For those of us involved in developing the project the challenge became to find many such tasks that stimulated mathematical thinking in its many guises.

Recently, back in New Zealand, a mathematics education colleague asked the question:

Imagine a quadrilateral with a square drawn on each side. Prove that the lines joining the centre of opposite squares intersect at right angles.

My immediate thought was:

John would use his lap-top, draw a quadrilateral with 2 fixed vertices and 2 variable ones. Construct the squares, join the centres, measure the angle, then move the variable vertices to show the invariance of the angle measure. Of course he would consider the general case and special cases such as concave quadrilaterals, the extreme case (a straight line, the 'crossed' quadrilateral (when two opposite sides intersect), and the alternative cases when the squares are drawn inside rather than outside the quadrilateral.

Being traditional I used coordinate geometry with each case and found with a few lines of algebra that the angle is indeed constant and measures 90°. Reflecting on this I noticed that John is more willing than I am to use technology, and rely on a visual thinking and a visual proof.

Noticing thinking

This focus on thinking again caused me to again question the nature of mathematics and to add the interrelated dimension thinking to knowing and doing. This focus was reinforced by numerous school curriculum initiatives in the first few years of the 21st century in which some countries introduced the 'competency' thinking into their curriculum documents. I immediately thought of the curriculum as being three-dimensional and envisaged knowing, doing and thinking on a Venn diagram with three overlapping sets. While the Venn diagram illustrates the overlapping relationship between these three strands, I prefer an alternative diagram, a triangular 'map' with three vertices (fig. 1) as it provides a focus for analysis.

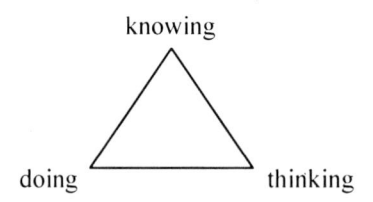

Figure 1. Knowing, doing, and thinking 'map'

Such a 'map' is useful for a teacher when preparing and reflecting on a learning task, it assists them in 'noticing' the balance between knowing, doing and thinking. I see such questions as the following being posed:
– Where is a particular learning task 'positioned' on this map?
– Can this task be modified to increase the emphasis on thinking?
– After using the task, did it fulfil my thinking (or doing or knowing) expectations?

This change in emphasis within mathematics to focus on thinking is important, but particularly so with the development of technology, as we now have tools for 'doing' mathematics and our emphasis needs to shift onto a 'thinking' in a conceptual rather than procedural way.

Thinking about thinking

The traditional way of thinking about thinking is to categorise thinking as critical (including reasoning and logic), creative, or meta-cognitive, but I do not find this useful. The two examples from geometry above show situations where logic is required although its form may vary, creativity is also involved, and meta-cognitive processes are needed as one monitors one's thinking through the various steps and stages of each task.

In addition, such a categorisation does not acknowledge the notion of embodied thinking or embodied knowing (Johnson, 1999) that involve at least three (I would say four) interrelated levels: biological, unconscious, phenomenological, and (for me) a conscious level. The biological level relates to our brain and neural networks (and to basic functioning of the body such as breathing and blood circulation); and the unconscious relates to bodily experience including our spatial and temporal orientations, our body movement, and to the way our senses process input. The phenomenological level is about how we experience our world. To these I would add the conscious level that involves not only critical, creative and meta-cognitive thinking, but also involves conscious aspects of the other three levels. For me the unconscious seems particularly important as much of early years learning including that of language, culture, and worldviews that influence us throughout our lives are learnt by unconscious imitation and by subtle absorption; and after the early years we continue to pick up messages from situations and people around us without actually thinking about how we develop these understandings.

A third way of considering knowing (and thinking) was offered by de Quincey (2005), he wrote of

... our "four gifts":

the Philosopher's Gift of *reason*;
the Scientist's gift of the *senses* (and methodology);
the Shaman's gift of participatory knowing through *feeling*; and
the Mystic's gift of sacred silence or *direct spiritual experience*. (pp. 1–2)

I think of these as *ways of knowing* or *ways of thinking* rather than as *gifts,* and I see them as ways that can be *learnt* rather than as *given*. I would add one more:

the Artist's way of *creativity*.

For me these five ways of knowing are an alternative way of considering most of the aspects of the four levels of embodied knowing or thinking.

Noticing as thinking

For me noticing is thinking. It occurs as we notice or bring to conscious awareness what we know through any way of working or experiencing, and as we analyse our behaviour and think about next time, that is, as we anticipate future possibilities. Such noticing occurs when we live 'mindfully' in a way in which our lived or phenomenological experiences are not simply experiences that happen but are also experiences that we are consciously aware of and accept, and may reflect upon in terms of possible similar future experiences.

An example of 'ethical action' illustrates this. An elderly woman at the supermarket is carrying a bag that splits and the contents fall out. Without thinking we stop and help her recover all her items. But then we might stop and consider, that is *notice*, what might we have done if this person had been a young man, or had had a very threatening appearance; where do we draw the line.

Similarly as teachers we can consider behavioural activities that we might wish to recognise and encourage, others that we might wish to anticipate and avoid, and subject (mathematical) activity that ideally might be encouraged or avoided. Such activity might be influenced by the teacher, the individual student, or classroom peers, and as teachers we need to consider the most desirable 'locus of control'. Such situations firstly require developing awareness or the skill of noticing, then giving consideration to the range of possible classroom strategies/interventions that might make the most of the situation.

Noticing and thinking about aims

It seems to me that mathematics (like other subjects) is always situated within a broader educational context, be it the home, a school, a university, or a work situation. In each situation the aims of the subject need to be seen as secondary to the more general aims within the context. In terms of work at school, in my early teaching career a report was published that outlined a set of aims for education that immediately appealed to me. I saw them as providing an important lead that moved teaching beyond merely a subject focus. Over the years I have often returned to these aims though I acknowledge that now I interpret them at a much deeper level than I did when I first read them. The aims were (Munro, 1969):
  the greatest emphasis should be put on fostering
  – the urge to enquire
  – the desire for self respect
  – a concern for others. (p. 1)

Now, nearly forty years later, a new school curriculum has been published in New Zealand (Ministry of Education, 2007), and, as in numerous other countries, it includes 'competencies' that can be considered as the aims for education. My immediate reaction was to notice the similarity between the 1969 aims and these competencies. They all seemed to map onto the three domains of growth that concern teachers. These three domains can be represented in different ways. If one sees them as mind, body, and society then the best representation might be a Venn diagram with three nested sets (and this can be extended with an additional set for the eco-system). However, the nested sets suggest if one works on the inner set then one is also working on the others. I prefer a further triangular map (fig. 2) with the intellectual, personal and social domains at the vertices.

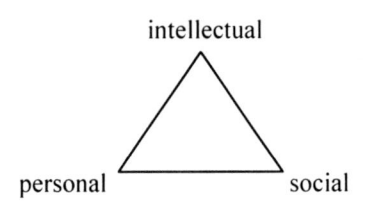

Figure 2. Domains of growth 'map'.

It is interesting to consider the tasks we use with our students in mathematics classes, other subjects, and extra-curriculum activities, and notice where they are and where they could be positioned on this map. Initially I mapped many mathematical tasks close to the intellectual vertex, but then realized that the classroom pedagogy usually involved personal and social growth. However, I was reminded of the saying that "primary school teachers teach children, high school teachers teach mathematics" and I wondered whether my pedagogical approaches facilitated personal and social growth as much as I hoped.

It is useful to use this 'map' from time to time to help us notice whether the tasks we use with our students concentrate too much on the intellectual domain at the expense of the other two; and to think of our responsibilities in terms of facilitating personal and social growth.

I see integrating the personal and social aspects of growth with intellectual growth as acknowledging the nested nature of mind/body/society. If we only emphasise the intellectual aspect then our subject is likely to be seen by many students as disconnected from them and society. However, how such integration might be achieved is a challenge.

CONCLUSION

In mathematics education there are many issues to be faced and changes we would like to see in our classrooms. These include: how do we encourage mathematical thinking, how do we encourage the personal and social development of our students as well as their intellectual development, and how do we encourage ways of working that may not be our own ways of working. I see the development of 'mindful awareness', or 'the discipline of noticing' as an important prerequisite step that underpins the changes we might wish to make.

REFERENCES

Begg, A. (2008). *Emerging curriculum.* Rotterdam, NL: Sense Publications.
Cockcroft, W. (Chair). (1982). *Mathematics counts* (Report of the Committee of Inquiry into the Teaching of Mathematics in Schools, DES&WO). London: Her Majesty's Stationery Office.
Davis, B. (1996). *Teaching mathematics: toward a sound alternative,* New York: Garland.
de Quincey, C. (2005). *Radical knowing: understanding consciousness through relationship.* Rochester, VT: Park Street Press.
Johnson, M. L. (1999). Embodied reason. In G. Weiss & H. F. Haber (Eds.), *Perspectives on embodiment: the intersection of nature and culture* (pp. 81–102). London: Routledge.
Maturana, H., & Varela, F. (1987). *The tree of knowledge: The biological roots of human understanding.* Boston, MA: Shambhala.
Ministry of Education. (2007). *The New Zealand curriculum.* Wellington, NZ: Learning Media/Ministry of Education.
National Council of Teachers of Mathematics. (1989). *Curriculum and evaluation standards for school mathematics.* Reston, VA: National Council of Teachers of Mathematics.
Munro, R. (1969). (Chair). *Education in change: Report of the curriculum review group of the New Zealand Post Primary Teachers' Association.* Auckland, NZ: Longman Paul (with NZPPTA).
Neill, A. S. (1962). *Summerhill: A radical approach to education.* London: Gollancz. (1968 ed., Harmondsworth, UK: Penguin.)

Varela, F. (1999). *Ethical know-how, action, wisdom, and cognition.* Stanford, CA: Stanford University Press.
Varela, F., Thompson, E., & Rosch, E. (1991). *The embodied mind: cognitive science and human experience,* Cambridge, MA: MIT Press.
von Glasersfeld, E. (1991). *Radical constructivism in mathematics education.* Dordrecht, NL: Kluwer.

Andy Begg
University of Auckland (New Zealand)

ANNE WATSON

17. THINKING MATHEMATICALLY, DISCIPLINED NOTICING AND STRUCTURES OF ATTENTION

THINKING MATHEMATICALLY

When I was a secondary mathematics teacher in the early 80s, Peter Gates introduced me to the now-classic *Thinking Mathematically*[1] in a professional development session. We worked on a problem of toasting multiple pieces of bread efficiently by employing various stacking and turning strategies (Mason, Burton & Stacey, 1982, p. 37). While this was interesting, and generated a need to think about representation of functions on a discrete domain, I did not think I was doing mathematics. I was applying known ideas but unable to see how the situation would extend my ideas of any of the concepts involved. Unless such tasks enriched or extended my conceptual understanding, perhaps by leading to a structure that I already recognised from another situation, I was not really interested. There was another problem too. I had always worked on mathematical problems by applying a range of strategies associated with suitable concepts until some transformation or representation happened to generate an expression or diagram that offered further transformational possibilities or insight into the mathematical structure of the problem. Often such insights came at inconvenient times: in the shower, or in the car, or early in the morning. Hadamard reports this phenomenon with reference to Poincaré (1945). I am by no means placing myself in their company but because the phenomenon applies to me I reasoned it might also apply for my students, given the opportunity to work on problems over time and to express insights. In Mason, Burton and Stacey's book, the processes of random, but informed, spattering of strategies and contemplating their effects, followed by mulling, are outlined but did not describe fully my experience of working on problems. For another, there was no detail about what you do to start, to get involved, to mull (p. 119), or how one strategy might remind you of another and lead to chains of reasoning, or even how sometimes those chains turn out to be circular or merely to rephrase the problem in many interesting but similarly intractable forms. The activities of manipulation, getting-a-sense-of, and articulating (p. 157) were meaningful ways to think about engaging with concepts, but *how* to manipulate and 'get a sense of' were not described.

Although working on the problems in the book was pleasurable, and the rubrics described in the book were authentic, and having them articulated was a step towards making them available as opportunities for learners, applying the book as a

S. Lerman and B. Davis (eds.), Mathematical Action & Structures of Noticing: Studies on John Mason's Contribution to Mathematics Education, 211–222.
© *2009 Sense Publishers. All rights reserved.*

teacher was difficult. There was a curriculum to be followed and I could not see, for a long time, how *Thinking Mathematically* could inform that process. However, I could see the potential for altering the nature of mathematics lessons so that mathematical thinking processes were the educational aim, and content knowledge could be seen as sensible and meaningful expression of the products of mathematical thought. As a teacher I could engage students with processes of conjecturing about generalities arising from inductive reasoning on examples. The slogans in the book: 'specialise, generalise, conjecture, convince' provided useful handles for these processes seen as empirically based. I could do little about proof, because the deductive reasoning associated with mathematical proof is a significantly different psychological process than the inductive reasoning that arises from seeing patterns in examples. However, the word 'specialise' implies selection of special cases rather than the sequential generation of examples – the latter having been picked up and used in schools as an investigative process. By contrast, exploration of the behaviour of special cases reveals structural behaviour in mathematical contexts and constraints on the domains of mathematical objects and relations. In the latter kind of investigation conjectures are likely to be about the nature of variables, the relationships between them, covariation, and theorems about connections and properties. In sequential exemplification conjectures are likely to be about generational formulae and functional relations.

These difficulties echo those explored by the developers of Realistic Mathematics Education: how can learners who have engaged fully with a situational problem then extract and extend a structural understanding of mathematics that can be further abstracted and used? In the context of using extended exploration and activities as the main form of teaching in 'mixed-ability' secondary classrooms I never managed to develop a coherent praxis about helping learners reason deductively and structurally, rather than inductively or pragmatically.

THE VERBS OF MATHEMATICS

The tensions I experienced with the ideas in *Thinking Mathematically* continued until I came across the existence of a list of 63 mathematical probes generated by Zygfryd Dyrslag in Anna Sierpinska's book (1994). Anna kindly translated these for me as I thought I was going to need them for my research on informal, integrated, assessment of students' mathematics. When the translation arrived, it was not what I thought it would be. Rather it turned out to be a list of some of the aspects of doing mathematics that I felt were missing both from Mason's and Pólya's work, probably because both of these were about solving problems, and also missing in my own teaching. The practices I saw arising from problem-solving approaches tended to look for answers in complex contexts rather than explore concepts. I contacted John Mason to discuss this list with him.[2] His immediate reaction was to recognise, like I had, that the list was an articulation of what was not usually described about the nitty gritty of doing mathematics at every level, whether it was directed towards solving problems, of whatever kind, or exploring new meanings, and he set about seeing how the list related to his fundamental beliefs about universal structures. My interest was more practical in that I needed intellectual tools to support task design, lesson planning, and classroom interactions both for myself as a teacher and also in my work as a teacher educator. Identifying the 'verbs' of doing mathematics (things I do when doing mathematics) as providing the raw material for constructing tasks (things I can ask learners to do) was a key moment in my work on teaching mathematics. It liberated me from an apparent choice between imitating techniques shown by a teacher or a book, and attacking complex problems for which one has to work out what procedures to apply. Articulation of the verbs of mathematics makes available many more kinds of question teachers can pose, whatever the overall shape of their lessons or curriculum goals. Furthermore, these questions can be posed about curriculum mathematics, so that mathematical thinking of the kind described in *Thinking Mathematically* can be applied to any and every mathematical topic, leading to extension and enrichment of conceptual understanding.

ACADEMIC EXPLORATION – NOTICING

This pattern of shared academic exploration that we developed to write *Questions and Prompts* (Watson & Mason, 1998) has continued, jointly and independently. We constantly:
- examine what it means to do, teach and learn mathematics;
- see particular instances as examples of generalities;
- describe these generalities;
- attempt to see how these new articulations relate to each other, and express them as structures of understanding;
- test these descriptions on teacher educator colleagues, teachers on professional development courses, and other researchers of mathematics teaching;
- disseminate.

Through this process, new ways of describing the detail of mathematical activity are made available in the field. There is a problem in that these descriptions become more and more detailed and are almost fractal in nature, so structure is necessary to handle them. However, any superimposed structure has to be meaningful and relate to people's own experiences of doing mathematics. It is likely that the reduction of 'mathematical thinking' to generic problem solving heuristics, or to generalisation from sequences of examples, as can be seen in many national curricula, may be due to a lack of personal mathematical experience of curriculum agents but could also be due to a lack of linguistic explicitness in the field. A phrase like 'mathematical thinking' can be mangled to mean 'modelling', 'application' and 'generalising patterns' by someone who has only ever been supported to explore maths in these situations. In some curriculum documents we have seen the phrase elaborated in terms of the taxonomy of Bloom - application, analysis, synthesis, evaluation – a taxonomy that captures nothing of the actions of doing and conceptualising mathematical ideas but merely focuses in the functional use of already-known ideas.

For this reason my sources of data are not what is *said* to happen in classrooms, but what I actually see. I use observations, and we both use our own mathematical activity, to identify new strategies, questions, tasks and interactive riffs which can be expressed as generalities. Another source of new thinking arises when a research seam somewhere else in the field suddenly appears to be related to this work in ways that were not apparent before. I shall use some of these seams to illustrate the process of synthesising theory and to demonstrate the emergence of further elaborations of mathematical thinking through what John would describe as the Discipline of Noticing (Mason, 2002). The main discipline is:[3]

− awareness to notice ideas in terms of how they relate to my current foci of attention;
− questioning the foci, our awareness, and the events that influence these;
− resonance with observed events in classrooms, and the utterances of colleagues;
− offering similar stimuli to others and observing their responses.

These aspects of human activity apply very well to doing mathematics, especially with others. Other authors in this collection show that they also apply to research, teaching, and professional development more generally.

THEORETICAL SEAMS

The seams are:

− Exemplification
− Conceptual understanding
− Variation and covariation
− Exercises in textbooks

Exemplification

One of the verbs of mathematics is 'exemplify' and 'example' is one of the types of mathematical object we identified in our original structuring of the 63 actions. We shifted our focus from the whole range of verbs used in *Questions and Prompts* to this one and had 'examples' as a background theme in all observation of and reflection on mathematical activity − our own separately and together, work with teachers and students, and work with colleagues in a variety of contexts. At first there were three sources of thought on which we drew: a review of research by Sowder (1980), Michener's work on understanding mathematics (1978) and Bateson's remarks about how learners needed to glimpse what was being exemplified by examples (1973). There had been a seminal paper by Mason and Pimm (1984) highlighting the role of generic examples in mathematics − Rowland (a student of Pimm) talked about this paper when describing how his primary teacher trainees understood, or misunderstood, the role of generic examples as proofs in elementary number theory (1998), and Bills (a student of Mason) extended the categorisation of examples to include particular and peculiar examples (1996). Meanwhile several Israeli mathematics educators had also been exploring the role and nature of examples and their methods of investigation had similar features to the pragmatic combination of experience, observation, reflection and structuring which we had developed. However, there was a key difference. The Israeli approach tended to follow Michener's assumption that example spaces are givens and the job of teachers is to give students access to these spaces through using canonical, representative and model examples. Our approach tended to be phenomenographic and hence focus on the personal example spaces on which teachers and students appeared to draw. The contrast between these approaches led to a realisation that the task of teaching might be seen as enabling learners to construct personal example spaces that might get closer and closer to those which would be generated from conventional mathematical definitions and methods. While this has much in common with Tall and Vinner's (1981) notion of concept images, it is, I find, more practical for teachers to think of example spaces. This image tells teachers that they need to think about what examples students have been exposed to, and to probe understanding by asking about particular examples, and to ask them to generate examples with particular properties. How learners create concept images with these is less directly tangible.

Conceptual understanding

There is something missing in current talk about mathematics teaching. Consider Kilpatrick et al.'s (2001) description of five strands of mathematics as a broadly agreed framework:
- procedural fluency
- conceptual understanding
- strategic competence
- adaptive reasoning
- productive disposition

Most curriculum change focuses on the last three of these, possibly with the inclusion of a critical dimension in countries who see mathematics as a tool for improving political consciousness. It is generally assumed that the first strand is pursued by traditional methods involving worked examples and practice from exercises. The conceptual strand is currently less explored in the English-medium literature and less can be assumed about it. It has become common to compare so-called traditional methods that focus on procedures with so-called reform methods that focus on the last three strands, but there is little consistent attention given in some mathematics education journals to modern work on how conceptual understanding develops within either of these approaches to teaching. While John's early work provided some detail to support the development of strategic competence and adaptive reasoning within mathematical contexts, and our joint work started by trying to provide more detail about teaching the specifically mathematical aspects of these, more recently we have become interested in analysing the affordances of mathematical tasks for *conceptualising* mathematics. In this work, the approach is again phenomenological: what experiences are likely to entice learners towards conceptualising mathematically?

The socio-cultural literature in mathematics education makes a good job of describing how mathematical practices can develop in classrooms, how mathematical activity develops through the discursive life of learners, how the lived experience of learners affects their development of identity in relation to mathematics, and how knowledge is structured by these aspects of classroom. We find, however, that at the heart of mathematical activity it matters what the task is, and we need descriptions of what practices, actions, and habits of mind are appropriate for the activity to be mathematics as understood by the wider field. For some of this we find ourselves in tune with Malcolm Swan (2008), who uses cognitive conflict as a principle to design tasks involving ordering, matching, classifying, including examples which challenge limited conceptualisations and hence provoke discussion and expression. He also offers multiple representations to be compared and related to each other. We also find we can make sense of some attempts to 'measure' task difficulty, such as the SOLO taxonomy, or cognitive load theory, because we look at the number of variables in a task and imagine how they might be manipulated to produce 'new' structures, or transformed to give new insights about relationships, or be controlled in order to make covariations more obvious. We also enjoy the detailed examination of exemplification which takes place in lesson study in some Asian contexts, and learning study in Sweden, in which choice of examples makes a difference to what learners will perceive, and hence what generalisations might become available for them. In these traditions, the conventional structures of mathematics are used as the basis for designing the objects presented to learners, and how they are presented, whatever the nature of the subsequent teaching.

Variation and covariation

In focusing on the role of examples in my own understanding of mathematical ideas, and in maintaining that focus in my work and my reading, I came across Variation Theory, which emanates from Ference Marton and colleagues (e.g., Marton & Booth, 1997). He is well known for, with Saljo, work on deep and surface learning. This distinction is not very useful in mathematics because of the multi-layered and contextual meanings of 'learning'. We need a more complex set of distinctions which can handle, for example, the 'depth' of understanding of number required to deal with infinitesimals and the 'depth' of understanding of number required to deal with surds – different kinds of 'depth' according to mathematical context. Variation theory offers a way to handle these different directions of 'depth' because it deals with dimensions of variation of concepts, rather than distinctions. It is therefore possible to explore along a dimension without looking for dichotomies or conflicts. It is also possible to explore a dimension in different ways depending on a learners' perception of the qualities of variation. For example, for understanding infinitesimals it is useful to explore single dimensional number in a continuous manner, iterating on ever decreasing intervals. For understanding surds it is more helpful to see number as fixed values about which we can never be accurate, so there is not a lot of point in homing in on them in ever-increasing magnification. Indeed, it might be more helpful to think of two kinds of number, those we do know exactly and those we have to treat algebraically. Thus for the dimension of variation 'real number' we have at least two different ranges of change depending on the context.

Here John's and my approaches differ; mine is to talk about 'ranges of change' as what the learner and teacher think of using; John prefers to use 'range-of-permissible-change' to describe what is mathematically meaningful – what changes can be made while maintaining mathematical meaning. His commitment to pure mathematics as a structured field of experience contrasts with what many may see as his post-modern commitment to what it means to be human – constantly in change, in process, an enactive organism whose views are only temporary and contingent. But the structures he offers are supports for making distinctions and being aware, not statements about what must be so.

For example, in the context of variation he talks about being aware of 'invariance amidst change' as a fundamental human state (Mason & Johnston-Wilder, 2004, p. 193). Sometimes we are not aware of what we think of as 'given' until something varies and suddenly we notice both what is static and what is dynamic. In Marton's view, we are only aware of what changes. In Piaget's view, there is a conflict to be resolved. In Vygotsky's view, an expert needs to direct the learner towards mathematical ways to perceive and make sense of an experience.

We explored the related idea of 'surprise' with a group of colleagues in the summer of 2007 (Watson & Mason, 2007). Our take on 'surprise' was that one could be intellectually surprised by the appearance of examples that conflicted with an idea we had constructed, or one could also be surprised by the appearance of examples that provoked realisation of something subconscious. In the workshop, we were reminded that surprise can also have a negative side for learners who want to feel secure and confident.

The main schools of thought about learning depend on difference as a stimulus for the development of ideas. Piaget's notion of cognitive conflict assumes a constant reorganisation in response to the impact of experience; Vygotsky talks of negotiating shifts from spontaneous everyday conceptualisation to the scientific concepts that are only met through formal educational settings. In each of these there is a sense of having to give up earlier mental commitments or organisations in order to accept new different ideas. Variation theory sees learning as the development of capabilities through discerning new variations and differences. It seems inevitable to us that, particularly in mathematics, new ideas have to involve the unpicking and reconstruction of old ideas, extending and enriching one's sense of mathematical structure. The problem of being gentle with learners' psychological reaction to surprise is not one we have worked on directly, because being human is to be changing, to be acting in situations, to be aware for and of new possibilities, to be constantly learning. The role of affect in mathematics is often taken to be negative, as anxiety or rejection of the whole subject, or as an injunction to generate enjoyment or fun. For us, affect is the feeling of doing mathematics and hence includes being puzzled, having puzzles resolved, becoming more capable, having 'aha!' moments. In my work, I have become convinced that these responses are possible for everyone with or without either general disaffection or the deliberate construction of 'enjoyment'.

Exercises in textbooks

On the face of it, this kind of work seems to be a long way from the problems posed in *Thinking Mathematically*, and also from the structures of awareness and self-noticing that John writes about. Indeed, visually the work on exemplification has more in common with fairly traditional textbooks than with what are currently regarded as 'good' teaching practices (Watson & Mason, 2006). This is not accidental. Algebra textbooks from the late-19th and early-20th century typically offer hundreds of 'practice' examples, but close examination of the variation within the examples reveals not merely a gradient of difficulty (often described as 'graded exercises') but a gradual unfolding of variation in a concept, encountered by doing examples that differ slightly from each other. In this way, a student who is bashing along being fairly successful at applying a limited rule of her/his own devising is confronted with an example which cannot be tackled in that way, and has to rethink. Students who see the questions as examples of enactment of a concept have their experience enriched and complexified by the exercise. Students who have only been taught 'tricks' come unstuck after a few straightforward examples. There is a fine balance to be attained between sequences of examples that invite inductive reasoning based on pattern and fluent repetition, and sequences that invite deduction about the underlying concepts.

The provision of sequences of examples in textbooks as a way to draw attentive students into understanding concepts has, in most textbooks, given way to rather random exercises that seem to be directed towards experiencing procedures for future test questions. We maintain that exercises can be constructed whose aim is conceptual understanding, rather than procedural fluency alone, and have seen examples in the publications of People's Press, Beijing, in textbooks by Bob Burn and John Backhouse, and in many old algebra texts.

STRUCTURE OF ATTENTION – LINKING MATHEMATICAL THINKING WITH NOTICING AND WITH MATHEMATICS

John's work on structures of attention, while referring mainly to doing mathematical work, applies equally well to professional development and empirical research (Mason 2002; 2003). Within a general model of how disciplined noticing informs our action he describes an enactive process of coming-to-know through manipulating mathematical objects; 'getting-a-sense-of' them; then articulating this sense. It is the middle part of this that most occupies our current explorations in mathematics. How do we 'get a sense of' mathematical ideas? To do this we think about plausible connections between senses and sense. While many thinkers see this connection as discursive, for us there are more immediate connections which can be conscious or subconscious, and which can take place both privately and with others. They can also be accidental or deliberate, self-directed or directed by others.

Different kinds of attention perform different functions in sense-making because they bring different aspects, and combinations of aspects, into the arena of sense-making. John describes these, in a typically terse manner, as
- Holding wholes;
- Discerning details (this not that, difference, classification, phenomenology, focus);
- Recognising relationships;
- Perceiving properties;
- Reasoning on the basis of properties (deducing from definitions).

The elaborations that follow are my own and not to be taken as John's. I use quadratic functions as an example space to illustrate attention.

Holding wholes

Those of us who are sighted can choose to look at a whole situation until choosing, or noticing, particular features of it for closer attention. For Wertheimer (1961) this process of gazing leads to foregrounding and backgrounding structures inherent in the object of attention. For John, this process provides an image that can be re-examined in more detail later. Presented with a quadratic graph I might notice its overall shape and direction; presented with an equation I might see a string of terms linked operationally. This kind of attention is not only descriptive, but also image-making.

Discerning details

The process of discernment is critical in mathematical understanding. It involves identification of what is there, describing the parts of an object in as much detail as we can, but being aware that by focusing on some details we may be ignoring others. It could be that the details that come to our attention are those that are changing, or those that remain stubbornly constant while other things change. I might speak about the graph, describing its shape from left to right across the image, or say where it crosses axes; I might say that the terms all involve x and some coefficients. Probably I would not think it important that the highest power is 2, because I have not been offered higher powers from which to distinguish '2' as being critical. A role of the teacher is to draw attention to details. This kind of attention can be descriptive and image-making, but can also be analytical.

Recognising relationships

Through perceiving variation and invariance we 'see' relationships between variables. Some variations show the extent of a concept because variation does not affect some stability elsewhere; some delineate the boundaries of what is possible because their variation causes other features to break down. We begin to think in terms of covariation, correlation, even to suspect causality. A collection of varying quadratic graphs (or a dynamic image) might lead me to recognise that however I vary the coefficients there is only one turning point, but that the number of places it crosses the x-axis varies. A set of quadratic equations all have 2 as their highest power for the independent variable, and 1 as the highest power of the dependent variable. By offering structured experience of possible variations, a teacher can lead students to pay attention to critical conceptual relationships. This kind of attention can be analytical, but can also go beyond to become conjectural and exploratory, and finally to the expression of relational functions as new objects.

Perceiving properties

Stabilities that persist even when there is variation elsewhere can be described as properties, and classes of objects can be defined as those that have particular properties. Quadratic functions have one turning point, may have two, one or no real roots, and have particular algebraic symbolic representations. Any object that has these features is a quadratic function, and all quadratic functions have these. From paying attention to examples and our experience of them in variation, we can understand general classes of related objects. Here conjecturing is very powerful, and so is the generation of definitions and classifications at different levels. Mathematical implications of properties can be conjectured.

Reasoning on the basis of properties/deducing from definitions

Now that we have defined a class of objects according to their common properties, our attention can shift to finding out what must be the case. The shift of attention is from inductive and abductive reasoning based on experience of details and how they relate, to deductive reasoning based on how certain properties relate within classes. The influence of Bourbaki can be seen clearly in this way of seeing mathematics, and it is one of several ways to point to the importance of helping learners become deductive reasoners. The van Hieles pointed to a similar shift in the context of geometrical understanding – that eventually we have to engage in rigorous formal reasoning (Burger & Shaughnessy, 1986). In Realistic Mathematics Education this shift is not described so markedly, but a related shift from generalising within a class of situations (horizontal mathematisation) to abstracting intellectual tools for use in other classes (vertical mathematisation) is described. Reasoning on this level can generate new objects and relations that might never be perceived except in the imagination.

BEYOND THE STRUCTURE OF ATTENTION

This structuring of attention indicates that teachers and learners always have a choice of what possibilities to pursue, what to bring into the public arena, how to direct attention to different ways of seeing objects and experiences, and hence afford different kinds of sense-making. The role of a knowledgeable teacher is critical in constructing sequences of examples, situations, and mediating devices which give students access to new ways of attending. In Vygotsky's terms, by doing this the teacher introduces the mathematical canon to students by constructing situations in which interaction between people enables the student to think is new-to-them ways that they could not have achieved on their own, or without experts.

Through intense focus on structures of mathematical attention, we therefore reconnect with the cultural take on doing and learning mathematics, and inevitably with the notion of zones of proximality.

This re-connection happened quite dramatically on a beach in Portugal, talking with Paola Valero about the problem of 'relevance'. John and I both felt that requiring mathematics to be 'relevant' to learners' current lives was limiting for learners and essentially trapped them into merely rehearsing their current situations. We saw the teacher's role partly as indicating new 'relevancies' that were beyond learners' current experience, so that they could imagine contexts beyond their own. We jokingly called slightly-beyond experience 'the zone of proximal relevance'. On reflection, we realised that the development in ZPD could be further unpacked, and that is mathematics teachers thought in terms of proximal zones of various kinds, then task design, example choice, and interactive riffs could always be directed towards such zones. Directing attention to a state slightly beyond current attention would be an example of this.

In typical manner, John took this idea and worked with it in a variety of ways: when doing mathematics, when teaching mathematics, when presenting mathematical tasks in various venues to a range of audiences, and in academic seminars. He draws on van der Veer and Valsiner's classification of zones of promoted action and of freedom of movement (1991), Gibson's ecological contextualisation of behaviour (Greeno, 1994) and his own intellectual commitment to seeing the human endeavour as combining behaviour, emotion and awareness. Action is promoted by the affordances of situations and the manipulation of variation within them. Freedom of movement is expressed by exercising range of change and is situationally constrained. Development is understood in terms of conscious awareness in and for oneself; emotion is understood as attunement and perception of relevance; behaviour as patterns of participation, including exercising mathematical habits of mind.

NOTES

[1] As I write a new, revised edition of *Thinking Mathematically* is being prepared.
[2] This is probably the moment to say, almost in Charlotte Bronte's words 'Reader, I (eventually) married him'.
[3] Discipline of Noticing is rather more complex than this; see Mason (2002).

REFERENCES

Bateson, G. (1973). *Steps to an ecology of mind*. London: Paladin.
Bills, L. (1996). The use of examples in the teaching and learning of mathematics. In L. Puig & A. Gutierrez (Eds.), *Proceedings of the 20th conference of the international group for the psychology of mathematics education* (Vol. 2, pp. 81–88). Valencia, SP.

Burger, W., & Shaughnessy, J. (1986). Characterizing the van Hiele Levels of development in geometry. *Journal for Research in Mathematics Education, 17*, 31–48.

Greeno, J. (1994). Gibson's affordances. *Psychological Review, 101*, 336–342.

Hadamard, J. (1945). *An essay on the psychology of invention in the mathematical field.* Princeton, NJ: Princeton University Press.

Kilpatrick, J., Swafford, J., & Findell, B. (Eds.). (2001). *Adding it up: Helping children learn mathematics.* Washington, DC: National Academy Press.

Marton, F., & Booth, S. (1997). *Learning and awareness.* Mahwah, NJ: Erlbaum.

Mason, J., & Johnston-Wilder, S. (2004). *Fundamental constructs in mathematics education.* London: Routledge.

Mason, J. (2002). *Researching your own practice: The discipline of noticing.* London: Routledge.

Mason, J. (2003). On the structure of attention in the learning of mathematics. *Australian Mathematics Teacher, 59*(4), 17–25.

Mason, J., & Pimm, D. (1984). Generic examples: Seeing the general in the particular. *Educational Studies in Mathematics, 15*(3), 277–289.

Mason, J., Burton, L., & Stacey, K. (1982). *Thinking mathematically.* Bristol, UK: Addison-Wesley.

Michener, E. (1978). Understanding understanding mathematics. *Cognitive Science, 2*, 361–383.

Rowland, T. (1998). Conviction, explanation and generic examples. In A. Olivier & K. Newstead (Eds.), *Proceedings of the 22nd conference of the international group for the psychology of mathematics education* (Vol. 4, pp. 65–72). Stellenbosch.

Sierpinska, A. (1994). *Understanding in mathematics.* London: Falmer.

Sowder, L. (1980). Concept and principle learning. In R. Shumway (Ed.), *Research in Mathematics Education* (pp. 244–285). Reston, VA: National Council of Teachers of Mathematics.

Swan, M. (2006). *Collaborative learning in mathematics: A challenge to our beliefs and practices.* London: National Institute of Adult Continuing Education.

Tall, D., & Vinner, S. (1981). Concept image and concept definition in mathematics with particular reference to limits and continuity. *Educational Studies in Mathematics, 12*(2), 151–169.

van der Veer, R., & Valsiner, J. (1991). *Understanding Vygotsky.* London: Blackwell.

Watson, A., & Mason, J. (1998). *Questions and prompts for mathematical thinking.* Derby, UK: Association of Teachers of Mathematics.

Watson, A., & Mason, J. (2006). Seeing an exercise as a single mathematical object: Using variation to structure sense-making. *Mathematical Thinking and Learning, 8*(2), 91–111.

Watson, A., & Mason, J. (2007). Surprise and inspiration. *Mathematics Teaching Incorporating Micromath, 200*, 4–5.

Wertheimer, M. (1945, 1961 enlarged ed.). *Productive thinking.* London: Tavistock.

Anne Watson
University of Oxford (United Kingdom)

CONTRIBUTORS AND ACKNOWLEDGMENTS

Rebecca Ambrose works with prospective and in-service teachers to enhance their mathematics instruction by exploring children's mathematical thinking. For the past six years she has been researching her work with teachers in high poverty schools in district-based professional development collaboratives in which the group discusses video clips of children from the district solving mathematics problems.

Andy Begg began his career teaching high school mathematics in New Zealand. He has worked in a number of universities, including two years at the Open University with John Mason; and is currently the postgraduate programme leader in the School of Education at Auckland University of Technology.

Laurinda Brown moved into mathematics teacher education at the University of Bristol, Graduate School of Education in the early 1990s, after teaching mathematics for 14 years at the same school in a department where students were encouraged to mathematise. Between school and university there was a short period of curriculum development as mathematics editor at the Resources for Learning Development Unit, which led to editing of *Mathematics Teaching* and, more recently, *For the Learning of Mathematics*. Research interests are related to the professional life of a mathematics teacher.

Encarnación Castro is a professor at the University of Granada in the area of Mathematics Education. She has a long professional career both in teaching and research on this area. Her research focuses on Algebraic and Numerical Thinking, but she has also studied students' mathematics performance in international tests and the cultural status of female mathematicians throughout history.

Enrique Castro belongs to the Spanish research Group "Numerical Thinking" and is a professor in the area of Mathematics education at the University of Granada. His main research interests are problem solving and representations. On these topics he has directed numerous studies and research projects financed by Spanish public research grants.

Olive Chapman is Professor and Assistant Dean at the Faculty of Education, University of Calgary. She teaches courses in secondary mathematics education. She is associate editor of the *Journal of Mathematics Teacher Education*. Her research interests and publications deal with mathematics teacher thinking, knowledge and learning; mathematical problem solving and word problems; students' sense-making of mathematics, and mathematics classroom discourse.

Alf Coles is an Assistant Headteacher at Kingsfield School, where he has worked for the last 12 years, starting as a teacher of Mathematics and then moving to Head of Mathematics. During this time he has engaged in a long-term research collaboration with Laurinda Brown looking into effective classroom practice and the development of classroom cultures. Alf is now exploring these issues in doing a part-time PhD at the University of Bristol Graduate School of Education. Before moving to Kingsfield, Alf taught for two years in London, a year in Eritrea and a year in Zimbabwe.

Brent Davis is Professor and David Robitaille Chair in Mathematics, Science, and Technology Education at the University of British Columbia. His research is developed around the educational relevance of developments in the cognitive and complexity sciences, and he teaches courses at the undergraduate and graduate levels in curriculum studies, mathematics education, and educational change. Davis has published books and articles in the areas of mathematics learning and teaching, curriculum theory, teacher education, and action research. He is currently editor of *For the Learning of Mathematics*.

Tommy Dreyfus earned a PhD in theoretical physics from the University of Geneva, Switzerland and is professor of Mathematics Education at the Department of Mathematics, Science and Technology Education of Tel Aviv University. His main research interests are processes of abstraction in mathematics learning, and students' and teachers' conceptions of proof. He has been associated with Educational Studies in Mathematics since 1990, most recently as Editor-in-Chief.

María José González is lecturer of Mathematics Education at the University of Cantabria (Spain). Her current research interest is secondary pre-service teacher training and integration of information technology in the mathematics curriculum.

Pedro Gómez is a researcher in the University of Granada (Spain) and a visiting lecturer in the University of los Andes (Colombia). His research interests focus on mathematics teachers training.

Anthony Harradine began teaching mathematics in 1984 after his father suggested he should become either a butcher or a teacher. From day one he became addicted to the dream of developing the best way to assist students to engage with this thing we call mathematics. Teaching has provided the opportunity for him to learn more mathematics, and more about how people see it, learn it and use it – for which he is very grateful. He is currently the Director of the Noel Baker Centre for School Mathematics at Prince Alfred College in Adelaide, Australia. Here he spends his time trying to share 25 years of ideas with anyone who will listen and dabbles in a little research too.

Dave Hewitt has taught in schools for 11 years before taking up a post at the University of Birmingham where he has now worked for 19 years and is a Senior Lecturer in Mathematics Education. His research interests form around the theme of the economic use of personal time and energy in the teaching and learning of mathematics. Particular avenues have included the teaching and learning of algebra and early number, as well as the development of several pieces of educational software.

Derek Holton officially retired as professor of pure mathematics at the University of Otago, NZ in 2009 to spend more time photographing wildlife but is now working without pay instead. His major interests as a mathematician were graph theory and combinatorics; as a maths educator his interests were problem solving, the more able students, and issues relating to mathematics enrolment at tertiary level. Derek really enjoy being in front of a class.

Steve Lerman is Head of Educational Research and Director of the Centre for Mathematics Education at London South Bank University. Steve was as a secondary school mathematics teacher for 15 years in London and in Israel before taking a PhD in mathematics education. Steve is a former President of PME and former Chair of the British Society for Research in Learning Mathematics. His current research draws on sociological and sociocultural theories in examining the learning and teaching of mathematics.

CONTRIBUTORS AND ACKNOWLEDGMENTS

Peter Liljedahl is an Assistant Professor of Mathematics Education in the Faculty of Education, an associate member in the Department of Mathematics, and co-director of the David Wheeler Institute for Research in Mathematics Education at Simon Fraser University in Vancouver, Canada. His research interests are focused on the professional growth of mathematics teachers in general, and the role that beliefs play in this growth. He is also interested in creativity, insight, and discovery in mathematics teaching and learning; mathematical problem solving; and numeracy.

Marta Molina is a young lecturer at the Mathematics Education department of the University of Granada (Spain). She teaches mathematics and mathematics education to prospective and in-service teachers. So far her research has focused on the integration of algebra in elementary education (Early-Algebra) and numerical thinking. She also co-launched and co-edits a Spanish research journal on mathematics education called PNA.

John Monaghan is Professor of Mathematics Education at the Centre for Studies in Mathematics Education, School of Education, University of Leeds, UK, where he enjoys teaching, supervising students and research. His research interests include students' understanding of calculus and of algebra, mathematical abstraction, linking school mathematics with out-of-school activities and the use of technology in the teaching and learning of mathematics.

Elena Nardi is Reader in Mathematics Education at the University of East Anglia in the UK. She studied mathematics in Thessaloniki (Greece) and mathematics education at Cambridge (MPhil) and Oxford (DPhil). Her research interests include the learning and teaching of mathematics at university level, student engagement with mathematics and the psychology of mathematical thinking. Her monograph *Amongst Mathematicians* was published by Springer in 2007. She is Joint Editor in Chief of *Research in Mathematics Education*, the official journal of the *British Society for Research into the Learning of Mathematics*, published by Routledge.

David Pimm worked for fifteen years at the Open University (1983–1997) as a colleague of John Mason's in the Centre for Mathematics Education. Since he moved to North America in late 1997, he has worked first at Michigan State University and, since 2000, at the University of Alberta. His research interests have consistently concerned the interrelationship between language and mathematics and its implications for teaching and learning.

Tim Rowland is a Senior Lecturer in Mathematics Education at the University of Cambridge. Having tried and failed at research in axiomatic set theory, he taught mathematics in schools with marginally more success before moving to teacher education and research in mathematics education. He is author of *The Pragmatics of Mathematics Education*, Falmer Press, and co-Editor of the Routledge journal *Research in Mathematics Education*.

Nathalie Sinclair is currently an assistant professor in the Faculty of Education at Simon Fraser University. Prior to that, she worked for almost four years at Michigan State University, cross-appointed between the Department of Mathematics and the Department of Teacher Education. Her current interests involve examining and understanding the role of aesthetics in mathematics thinking and learning, especially in light of recent theories of embodied cognition. Nathalie has also focused her research on the uses and implications of expressive technologies such as dynamic geometry software.

CONTRIBUTORS AND ACKNOWLEDGMENTS

Elke Söbbeke has been teacher for primary school and completed her PhD thesis at the University of Dortmund. Her supervisor was Prof. Dr. Heinz Steinbring. Since 2005 Elke Söbbeke has been a researcher in mathematics education at the University of Duisburg-Essen. Her scientific interests include the scientific foundations of mathematics education, research on curricula of school mathematics, and the analysis of the specific epistemology of mathematical knowledge in the context of mathematical representations.

Heinz Steinbring has been professor of mathematics education at the University of Duisburg-Essen since 2004. Before that he was professor of mathematics education (1993–2004) at the University of Dortmund and a researcher in mathematics education at the Institute for the Didactics of Mathematics (IDM, University of Bielefeld, 1974–1993). His scientific interests include the scientific foundations of mathematics education, research on curricula of primary and lower secondary school mathematics, and the analysis of the specific epistemology of mathematical knowledge in classroom interaction.

David Tall is Emeritus Professor of Mathematical Thinking at the University of Warwick, having obtained a first class degree and a DPhil in mathematics at Oxford prior to three years at Sussex University and forty years at Warwick where he began as a lecturer in mathematics with special interests in education. He obtained a second PhD in Education with Richard Skemp. His major theoretical work is based on empirical research with colleagues and PhD students including the notion of concept image with Shlomo Vinner, procept with Eddie Gray and the subsequent theory of three worlds of mathematics.

Mike Thomas is an associate professor in the mathematics department of The University of Auckland, New Zealand. His research field is mathematics education, particularly the use of technology in teaching and learning, the teaching and learning of algebra and calculus, advanced mathematical thinking, teacher knowledge, and the relationship of cognitive neuroscience to mathematical thinking and use of gestures.

Anne Watson Anne Watson is Professor of Mathematics Education at the University of Oxford, having taught mathematics in English secondary schools for 14 years before becoming a teacher educator. She is currently interested in how teachers construct mathematically challenging tasks for all students, and how ongoing engagement with mathematics enriches teaching. She and John were married in 2000, and together they hold mathematics evenings in their home, and run an annual four-day workshop on mathematics pedagogy.

Rina Zazkis is a Professor of Mathematics Education in the Faculty of Education, an associate member in the Department of Mathematics, and co-director of the David Wheeler Institute for Research in Mathematics Education at Simon Fraser University in Vancouver, Canada. Her research is in the area of undergraduate Mathematics Education, with a general focus on mathematical content knowledge of prospective teachers and the ways in which this knowledge is acquired and modified. Teaching, learning and understanding elementary Number Theory has been a specific focus of her studies.

ACKNOWLEDGMENTS

In addition to thanking the authors for their contributions, the editors would like to express appreciation to **Steven Khan** and **Lissa D'Amour** for their assistance with proofreading, formatting, and indexing.

AUTHOR INDEX

Adams, V., 7
Ainley, J., 186
Ambrose, R., 88, 121–133, 223
Archimedes, 192–193, 198
Aristotle, 115
Artigue, M., 45, 119
Arzarello, F., 109
Ashlock, R. B., 168
ATM, 4, 135, 159
Ausubel, D., 6

Bachelard, G., 4
Backhouse, J., 218
Baker, J., 3
Balacheff, N., 4, 199
Bass, H., 176
Bateson, G., 156, 214
Begg, A., 163, 203–209, 223
Behr, M., 133
Bennett, J. G., 3–6, 10, 12, 149
Benoist, C., 61
Benson, S. R., 64
Bills, L., 122, 214
Biza, I., 175
Blanton, M. L., 121
Bloom, B., 21
Booth, L. R., 99, 216
Borghart, I., 63
Bosch, M., 109
Bourbaki, 2, 220
Briggs, L. S., 180
Brizuela, B. M., 77
Bronte, C., 221
Brown, L., 88, 92, 136, 140, 141, 147–159, 223
Brousseau, G., 4, 8, 115
Bruner, J., 137, 144
Burger, W., 220
Burn, R., 49, 218
Burton, L., 4, 17, 19, 31–32, 59, 65, 70, 211
Calvino, I., 10, 13
Campbell, S., 169
Carpenter, T., 122
Carraher, D., 77, 95
Castro, E., 88, 121–133, 223
Castro, E., 88, 121–133, 223
Chapman, O., 17, 63–73, 223
Charalambous, C. Y., 174
Chevallard, Y., 4, 111
Chinnappan, M., 45
Christianidis, J., 193–194
Clandinin, D. J., 120
Clark, L. M., 174

Cockcroft, W., 204
Coles, A., 88, 135–144, 150, 156, 223
Connelly, F. M., 120
Cooney, T., 11
Cooper, T. S., 75–76, 78
Courant, R., 192
Csikszenhmihalyi, M., 138

D'Amour, L., 226
Davis, B., vii–viii, 138, 155, 157, 204, 223
Davies, B., 90
Davydov, V., 87, 101, 104–109
de Corte, E., 63, 70
de Quincey, C., 207
Devlin, K., 75
Dewey, J., 136–137
Dienes, Z., 6, 107
Diophantus, 193–194, 196
Dreyfus, T., 78, 87, 101–109, 224
Drijvers, P., 45, 103
Drury, H., 122
Dyrslag, Z., 212
Earnest, D., 77
Eliot, T. S., 159, 198
Engels, F., 101
Erlwanger, S., 11, 133
Euclid, 192

Farrell, M. A., 180
Fauvel, J., 190–191
Findell, B., 63
Franke, M., 122
Freud, S., 22
Freudenthal, H., 6, 102

Gadamer, H.-G., 155
Gagne, R. M., 180
Gates, P., 12, 211
Gattegno, C., 6, 8, 9, 11, 58, 87–93, 99, 135–138, 147–150, 198
Geramita, J., 198
Ghousseini, H. L., 174
Gomez, P., 163, 179–187
González, M. J., 163, 179–187, 224
Gowar, M., 121
Graham, A., 76, 121
Gray, E., 27, 190–191
Greeno, J., 221
Greer, B., 70
Griffin, P., 12
Guin, D., 45
Gurdjieff, G. I., 10, 149

AUTHOR INDEX

Hadamard, J., 211
Haggarty, L., 203
Hall, D., 198
Hall, R., 51, 53
Hansen, A., 186
Hardy, G. H., 51
Harradine, A., 17, 31–47, 224
Haseman, K., 99
Hatch, G., 49
Healy, H., 59
Heaney, S., 199
Heidegger, M., 135
Heisenberg, W., 9
Hejny, M., 122
Hershkowitz, R., 102–106
Hewitt, D., 87, 89–100, 121, 224
Hilbert, D., 192
Holton, D., 17, 31–47, 224
Hong, Y. Y., 45
Hoyles, C., 59
Hughes, T., 199

James, W., 11
Jaworski, B., 5, 10, 87, 111
Jirotkova, D., 122
Johnson, M. L., 206
Johnston-Wilder, S., 12, 76, 179, 217
Joseph, G., 190

Kaput, J., 121
Keats, J., 61
Khan, S., 226
Kidron, I., 102
Kieran, C., 45, 103, 121
Kilpatrick, J., 63, 107, 215
Kratochvilova, J., 122

Lagrange, J. B., 45
Langer, S., 1
Lannin, J., 82
Lee, H. S., 64
Lenfant, A., 109
Lerman, S., vii–viii, 224
Leron, U., 192
Lester, F., 31
Levi, L., 122
Liljedahl, P., 163, 165–176, 225

Mahapatra, J., 200
Marton, F., 216
Merton, T., 147
Mascaró, J., 9
Maslow, A., 4
Mason, J. H., i–viii, 1–222
Maturana, H., 6, 204
Max-Planck-Gesellschaft, 77
McLeod, D., 7

Michener, E., 214, 215
Molina, M., 88, 121–133, 225
Monaghan, J., 87, 101–109, 225
Morris, A. K., 77
Mousley, J., 180
Munro, R., 208

Nardi, E., 87, 111–120, 175, 225
NCTM, 63, 70
Netz, R., 192–193
Neill, A. S., 203
New Zealand Ministry of Education, 208
Newton, I., 116
Nicolet, J., 9
Nichols, E., 133
Nørretranders, T., 11

O' Driscoll, D., 199
Orage, A., 10
O' Reilly, M., 155–156
Ouspensky, P. D., 149
Ozmantar, M. F., 104, 105
Page, J., 198
Palmer, P., 155
Papy, F., 6
Piaget, J., 6, 186, 217
Pimm, D., 10, 56, 121, 163, 189–202, 214, 225
Poincaré, H., 7, 211
Polyá, G., 2–4, 6, 45, 50, 70, 171, 212
Pratt, D., 186
Prediger, S., 109
Presmeg, N., 112
Proulx, J., 196

Rabardel, P., 45
Radhakrishnan, S., 9
Raymond, L., 9
Robbins Report, 49
Rosch, E., 204
Rotman, B., 197
Rowland, T., 17, 49–61, 214, 225
Ruthven, K., 150

Säljö, R., 216
Schliemann, A. D., 76
Schön, D., 111
Schroeder, T. L., 31
Schwarz, B. B., 102, 105
Sealy, J. T., 174
Sfard, A., 121
Shaughnessy, J., 220
Sierpinska, A., 212
Simon, M., 180, 186
Silver, E. A., 175
Sinclair, N., 163, 165–176, 189–202, 225
Skemp, R., 6, 20–23, 107
Söbbeke, E., 17, 75–83, 226

AUTHOR INDEX

Sowder, L., 214
Spence, B., 88
Spence, M., 46, 121–123, 127, 132
Staats, S., 191
Stacey, K., 4, 17, 19, 31–32, 59, 65, 70, 211
Steele, G., 1–2
Steinbring, H., 17, 75–83, 226
Stephens, M., 122
Stern, E., 77
Storm, H., 13
Sumara, D., 155–157
Swafford, 63
Swan, M., 216
Szydlik, J. E., 64
Szydlik, S. D., 64

Tahta, D., 4, 197
Tall, D., 17, 19–29, 76, 215, 226
Taplin, M., 63
Thera, N., 138
Thomas, M., 17, 31–47, 76, 226
Thomas, M. O. J., 45
Thompson, A., 3, 7
Thompson, E., 204
Thompson, P. W., 183
Tzur, R., 186

Valero, P., 221
Valsiner, J., 221
van der Akker, J., 111
van der Veer, R., 221
van Hiele, 25, 220
Varela, F., 147–150, 204
Vergnaud, J., 166
Vérillon, P., 45
Verschaffel, L., 63, 70
Vinner, S., 215
Vlassis, J., 27
von Glasersfeld, E., 4, 204
Vygotsky, L. S., 217, 220

Wager, W. W., 180
Warren, E., 75–76, 78
Watson, A., *viii*, 50, 95, 164, 168, 175, 196, 211–221, 226
Wertheimer, M., 219

Yusof, Y. b. M., 28

Zachariades, T., 175
Zazkis, R., 163, 165–176
Zwicky, J., 199–200

SUBJECT INDEX

A
abstraction, 2, 9, 17, 75–79, 81, 82, 87, 92, 95, 101–109, 121, 137, 148, 155, 156, 186, 189, 199, 212, 220
action research, 111
aesthetics, 25, 88, 135–144, 189, 198
algebra, 7, 24, 26–29, 36, 38, 44, 47, 49, 51, 53, 54, 59, 75–76, 88, 96, 103, 121–133, 148, 155, 189, 194, 196, 197, 205, 217, 218, 220
arbitrary and necessary, 90
arithmetic, 24, 26, 27, 29, 63, 75–77, 97, 121–126, 128, 132, 189, 190, 193, 194, 197, 198
assessment, 8, 49, 165, 187, 212
Association of teachers of mathematics (ATM), 4, 99, 135, 159
attention, 3, 5, 7, 8, 12, 24, 25, 35, 45, 46, 50, 88, 90–92, 100, 112, 119, 122, 125–127, 136–138, 144, 149, 153–155, 164, 187, 192, 193, 196, 199, 211–221. *See also* shifts of attention and structures of attention
of chair, 144
free, 128, 129, 132
shifts of, 23, 25, 26, 101–109, 127, 132, 156–157, 165–176, 220
awareness, 3, 4, 12, 13, 17, 49, 58, 60, 67, 73, 75, 88–100, 111, 118, 123, 136, 142–144, 148, 150, 155, 158, 159, 163, 164, 167, 168, 170, 172–176, 179, 190, 198, 207, 214
educating, 8, 12, 90–92, 98, 138, 163
structured, 9–10
types of, 166

B
being mathematical, 2, 156
Bhagavad Gita, 9
Bloom's taxonomy, 21, 213
BODMAS, 27

C
caring, 3, 20, 137
change, 5, 10, 20, 25, 27–29, 43, 45, 60, 63, 64, 71, 72, 76, 78–80, 91, 93, 102, 107, 117, 118, 122, 125, 127, 130, 131, 135, 137, 142, 143, 151, 156, 163, 165, 187, 196, 197, 203, 206, 209, 215, 217, 219, 221
Cockroft report, 204
computation
habit and skill, 123, 129
conjecturing, 3, 6, 17, 31, 32, 36–40, 46, 47, 50, 52, 53, 56, 58–60, 64, 92, 111, 129, 141, 143, 149, 154, 156, 212, 220
consciousness, 11, 115, 118, 119, 150, 159, 166, 186, 192, 206, 207, 215, 218, 221

consolidation, 1, 103, 105
constructivism, 4, 204
creative process, 31, 33, 46, 47
cultures of generality, 189–202
arithmetic, 189, 190, 193, 194, 197, 198
dynamic geometric, 194–197
geometric, 189, 193–197, 200
poetic, 189, 191, 198–199
curriculum, 1, 49, 76, 87, 89–100, 121, 166, 173, 175, 204, 206, 208, 212, 213, 215
design, 179
formal, 1, 100

D
deduction, 2, 26, 49, 59, 60, 212, 218, 220
definition
shift from description, 25, 26, 29
design-based research, 123–125
didactique, 4, 111
discipline of noticing, 3, 11, 23, 87–88, 135, 136, 149, 203, 209, 214, 221
distinction making, 7, 9, 10, 148, 158, 217

E
early learning, 76–77, 121–133
embodiment, 24–27, 29, 73, 136, 139, 206, 207
emotion, 8, 9, 20–22, 28, 68, 149, 150, 159, 179, 221
enquiry, 3, 4, 9, 11, 12, 60, 137, 208
teacher, 5–6
example, 1, 21, 31, 49, 63, 75–83, 90, 102, 111–119, 122, 137, 147–159, 167, 184, 190, 204, 212
spaces, 153, 168, 169, 196, 215, 219
experiential learning, 65–67, 82
experimenting, 12, 21, 31, 33–35, 38, 46, 59, 123–131
explanation
levels of, 20, 21
extra-spection, 7. *See also* spection

F
feedback, 64, 113, 118, 152, 175
Fibonacci, 55–57
FOIL, 27
formalism, 24
fractions, 1, 7, 11, 75, 89, 94–100, 102, 119, 166–168

G
generalizing, 3, 6, 24, 27, 31, 38, 41, 47, 56, 59, 60, 75–83, 88, 92, 95, 107–108, 117, 121,

SUBJECT INDEX

130, 133n8, 140, 142, 144, 148, 152, 155, 190, 192–194, 197–199, 205, 212, 213, 216
geometry, 1, 3, 23–25, 28, 29, 33, 49, 53, 90, 189, 194–197, 205, 206

H
Heisenberg principle, 9
hyperbolic function, 157–159

I
induction, 56, 60, 107, 116
inner research, 111–120
instrumental genesis, 45
intention, 5, 6, 9, 21, 51, 103, 108, 123, 137, 153, 159, 171–173, 175, 179, 190, 203
International commission on mathematical instruction (ICME), 4, 176
intra-spection, 8. *See also* spection
intro-spection, 8, 66, 67. *See also* spection
intuition, 113, 115, 116, 135, 154

K
knowing
    to act, 121–133
    types of, 46, 122, 130, 166, 174
knowledge
    types, 121–134

L
labelling, 10, 11, 50, 66, 67, 81, 90, 93, 94, 123, 141, 144, 151
language, 2, 5, 24, 60, 75, 90, 93, 95, 118, 124, 136, 152, 155, 156, 169–170, 172, 173, 175, 189, 198, 207. *See also* symbolism and labelling
lesson
    explaining, 175
    goals, 165, 175, 187, 213
    paths, 182–187
    planning, 70, 72, 148, 155, 165, 167, 180, 187, 213
    play, 163, 165–176
listening, 17, 132, 138, 144, 147, 155, 156, 200, 204
lived experience, 9, 10, 66, 216

M
mathematical practices, 3, 115, 193, 216
mathematics education
    critiques, 10, 12
    evolution of field
    research foci, 7, 8, 11, 31, 66, 67, 111, 119, 120n, 121, 176
mathematics-for-teaching, 174
matrix algebra, 150–153
metacommenting, 144, 156
metaphor, 94, 96, 122, 132, 199

method, 3, 8, 11, 25, 27, 29, 32, 40, 63, 64, 70–71, 75, 77, 79, 82, 103–105, 116, 120n, 121, 129, 137–143, 165, 166, 175, 179, 181, 189, 190, 193, 194, 198, 207, 214, 215

N
narrative inquiry, 120
noticing. *See* discipline of noticing

O
ogden, 90, 92

P
Pascal's triangle, 56, 57
patterns, 34–36, 75–77, 82, 95, 121, 124, 126, 127, 138, 148, 170, 173, 186, 212, 213, 221
    in research, 138
pedagogical technology knowledge (PTK), 45
perception, 21, 23–25, 29, 97, 136, 148, 150, 159, 163, 196, 216, 221
phenomenography, 215
phenomenology, 8, 9, 79, 206, 207, 215, 219
poetry, 189, 198, 200
problem solving, 4–7, 17, 19–29, 31, 32, 41, 45, 46, 158, 171, 212, 213
    teaching and, 63–73
procedural knowledge, 24
procept, 24, 26, 27
professional development, 4, 5, 13, 175, 204, 211, 213, 214, 218
proof, 17, 19–29, 31, 32, 37–43, 47, 53–60, 61n3, 113, 116, 120n, 156, 192, 193, 198, 199, 205, 212, 214
    anxiety and, 22–28
    non-algebraic, 53–55
    non-routine, 65, 70
    purposes, 59
psyche, 7, 9, 148
Pythagorean triples, 34, 44, 52

R
RBC model, 105, 106
Realistic Mathematics Education, 212, 220
reasoning, 49, 50, 58, 60, 80, 92, 120n, 168, 206, 211, 212, 215, 218–220
reflective abstraction, 186
reflective practice, 111, 175
relational thinking, 122–132
relationships, 9, 10, 20, 21, 24–28, 35, 36, 45, 51, 53, 63, 89, 104–108, 111, 112, 119, 120n, 122, 124, 130, 132, 158, 183, 194, 195, 204, 206, 212, 216, 219–220
representation, 57, 77–82, 102, 113, 115, 180, 185, 197, 208, 211, 216
    types of, 78, 82, 112, 124, 193, 195, 196, 220
Robbin's report, 49

S

schemas, 23
self-study, 63–73
shifts of attention, 102, 103, 108, 156–157, 165–176
situatedness, 12, 102
socio-cultural, 5, 7, 216
specializing, 3, 6, 31, 59, 60, 92, 192, 212
spection, 119
 extra-, 7
 intra-, 8
 intro-, 8, 66, 67
spirituality, 88, 147–159
standards
 National council of teachers of mathematics (NCTM), 204
structuralism, 2
structures of attention, 211–221
symbolism, 24, 26, 29, 121, 124
systematics, 10, 12, 66, 67, 87, 111, 119

T

task analysis, 183
teacher education, 7, 49, 63, 64, 72–73, 111, 150, 157, 165, 176
teaching, 1–3, 7–9, 17, 28, 31, 32, 45, 49, 59, 63–73, 76, 77, 82, 87–89, 92, 111–121, 123–132, 135–137, 141, 142, 144, 147–159, 163, 165–176, 179–187, 189, 198, 204, 208, 212–216, 218, 221
technology, 45, 47$n1$, 64, 115, 197, 205, 206
 computer based, 189
 pedagogical knowledge, 45, 197
theory, 1–13, 17, 21–23, 45, 56, 58, 61$n5$, 67, 70, 105, 115, 116, 118–119, 120n, 170, 172, 173, 204, 214, 216, 217
 practice *versus*, 1
Thinking Mathematically, 4, 17, 21–23, 28–29, 31, 65, 87, 88, 144, 156, 211–221

U

Upanishads, 9

V

van Hiele levels, 25
variation, 29, 59, 75, 95, 163–164, 169, 170, 191, 192, 203, 214, 216–221
 co-variation, 212, 214, 216–219
variation theory, 216, 217
verbalization, 25, 105, 121, 124
visualization, 75–83, 112–117